Contributions from Biology Education Research

Fayetteville, USA
Cynthia Passmore, University of California Davis
Davis, USA
Marcus Grace, University of Southampton
Southampton, UK
Marie Christine Knippels, Utrecht University
Utrecht, The Netherlands

Contributions from Biology Education Research (CBER) is the international book series of the European Researchers in Didactics of Biology (ERIDOB). The series includes edited collections of state-of-the-art papers presented at the ERIDOB international conferences, and monographs or edited collections of chapters by leading international scholars of the domain. The aim of the series is to shed light on global issues and trends in the teaching and learning of biology by gathering cutting edge research findings, theoretical views, and implications or concrete suggestions for everyday school practice regarding biology. The books may serve as resources for (a) getting informed about the most recent findings of biology education research to possibly integrate them in new personal research, and (b) studying about the teaching and learning of biology as a pre-service or in-service biology teacher. So, the main audiences for the series range from senior to early career biology education researchers and pre- or in-service biology teachers working at all educational levels. Book proposals for this series may be submitted to the Publishing Editor: Claudia Acuna E-mail: Claudia.Acuna@springer.com

Dina Tsybulsky • Tali Tal
Editors

Citizen Science Contributions to Biology and Environmental Education

Realities, Impact and Opportunities

Springer

Editors
Dina Tsybulsky
Faculty of Education in Science and Technology
Technion Israel Institute of Technology
Haifa, Israel

Tali Tal
Faculty of Education in Science and Technology
Technion Israel Institute of Technology, Samuel Neaman Institute
Haifa, Israel

ISSN 2662-2319 ISSN 2662-2327 (electronic)
Contributions from Biology Education Research
ISBN 978-3-031-99297-1 ISBN 978-3-031-99298-8 (eBook)
https://doi.org/10.1007/978-3-031-99298-8

This Springer imprint is published by the registered company Springer Nature Switzerland AG.
The registered company address is: Gewerbestrasse 11, 6330 Cham, Switzerland

Dedicated to the cherished memory of Tal Shavit.

Contents

Editors and Contributors

About the Editors

Heidi L. Ballard is a Professor of Environmental Science Education and Founder and Faculty Director of the Center for Community and Citizen Science (CCS) in the School of Education at the University of California, Davis. She conducts research focused on science and environmental learning that link communities, scientists, and environmental action. Her work focuses on what and how adults and youth learn through community science, citizen science, and other forms of public participation in scientific research. She works in varied contexts, from natural history museums to Uganda farmer field schools to carceral settings, using primarily qualitative research methods in partnership with teachers, environmental educators, research scientists, and community-based organizations. Her publications span the fields of environmental and science education, ecology, conservation, social science, forestry, and agriculture. She was a high school biology teacher and science curriculum developer in California, before earning her MA in Science Education and Teaching Botany, and Ph.D. in Environmental Science, Policy and Management focused on forest ecology and natural resource social science. She trains undergraduates, pre-service teachers, and graduate students across the university in science education and participatory action research methods.

Yael Bar Nes was born in Jerusalem and educated in a "Bnei Akiva" Ulpana in Israel. She received her BSc degree from the Faculty of Agriculture, Food and Environment of the Hebrew University. She received a teaching credential by the organization of Teach First Israel and became a science and technology teacher in high schools. Yael studied, taught, and participated in research about the Multi-Faceted Holistic Approach (MuFHA) of science teaching. She received the M.Ed. degree from the Technologies in Education Graduate Program at the Faculty of Education, University of Haifa. As a researcher in the Taking Citizen Science to School (TCSS) research center, she designed and led the "Bermuda Buttercup

Mystery" citizen science project. This project was the study field of her master's research focusing on fostering Environmental Science Agency in citizen science contexts.

Philippe Bastiaenssens is a lecturer in physics, natural sciences, and STEM didactics at the Karel de Grote University of Applied Sciences and Arts (Belgium). At the research center Future-Driven Education, he studies science and STEM education, citizen science in education, and student participation.

Haya Ben Simon is a doctoral candidate [direct Ph.D. track] in the Faculty of Education in Science and Technology at the Technion. She graduated from the Technion with a B.Sc. degree in Biology (2013) and a B.Sc. degree in Biology Education (2016). Haya has been a Biology, Science, and Biotechnology high school teacher for the last seven years. Her research focuses on the development and integration of Scientific Reasoning, Nature of Science Understanding, and Informal Statistical Reasoning through student engagement in Citizen Science.

Angela Bermejo graduated in Biology in 2019 and is a teacher of Compulsory Secondary Education and Baccalaureate. Since 2022, PhD student at the Universidad Autónoma de Madrid in the branch of Didactics of Experimental Sciences. Her line of research is based on the use of innovative proposals, such as board games, to bring science closer to students and improve the quality of education. She has designed several educational games for the didactics of environmental issues and emergency disease outbreaks. She has also participated in several teaching innovation projects and, recently, in the citizen science project "Flebocollect" in which STEM activities are integrated.

Jelle Boeve-de Pauw is an assistant professor of science education at the Freudenthal Institute (Utrecht University). As a biologist and educational scientist, he is interested in environmental education, education for sustainable development, science and STEM education and citizenship education. Issues of measurement and evaluation in these fields are central to his research, and he has published extensively on the subjects. He has an MSc in Biology (2003) and a PhD in Education (2011) from the University of Antwerp in Belgium and was awarded the 2014 Rudi Verheyen award for best research with relevance for environmental policy in Flanders. Before moving into academia Jelle has worked as a museologist, science educator, and communicator at the Royal Belgian Institute for Natural Sciences. Currently, at Utrecht University he teaches in the master program Science Education and Communication, with a focus on subjects relating to science-society dynamics and research methods in the educational sciences. He is also affiliated to Karel de Grote University of Applied Sciences and Arts (Belgium) as a research project leader and serves on the academic committee of Eridob, the European Researchers in Didactics of Biology. Jelle holds a position as associate editor for the Springer journal "Environment, Development and Sustainability," for "Education and Public Awareness" section.

Tarcísio Milagres PhD, is a biologist and parasitologist involved in education and the generation of relevant knowledge for promoting socio-environmental impact. He has participated in interdisciplinary projects in Brazil, the United States, Spain, and Portugal. Based on constant innovation, he combines his scientific background with his communication and empathy skills to implement effective strategies for project development with a critical and responsible approach.

Bart de Groot is a chemist and graduate student at Utrecht University's Master programme in Science Education and Communication. He has a keen interest in communicating science with the public and in facilitating the learning of science through fostering participation in authentic research. His research focuses on the needs that science teachers in secondary education experience when implementing citizen science projects into their teaching.

Mart Doms is a senior researcher at the Karel de Grote University of Applied Sciences and Arts (Belgium) within the research center Future-driven Education. Mart's areas of interest include education for sustainable development, science and STEM education, citizen science in education and authentic learning environments with which she works on a daily basis both within her role as responsible for the "Authentic Learning" research line and as a primary school teacher trainer. She has an MSc in Education (2019) from the University of Antwerp in Belgium.

Michal Dvir is a lecturer in Mathematics and Statistics Education at the University of Haifa. Her research focuses on raising the accessibility of complicated mathematical and statistical practices to young learners. Her work includes developing epistemological models for central practices or ideas and, based on these, designing and researching informal adaptations, often incorporating mathematical or statistical modeling-centered pedagogies. Her research also examines how in-service and pre-service mathematics teachers can be introduced to these pedagogies. She has a BSc in Mathematics, and her MA and PhD were in Mathematics Education.

Amram Eshel was born and raised in Israel, and received his academic degrees from Tel-Aviv University, where he was a member of the academic staff since 1980 until his retirement in 2012. During those years, he held short visiting appointments at Ohio Agricultural Research Center in Wooster, Ohio, Michigan State University, Penn State, CNRS at Gif-Sur-Yvette, and University of Western Australia. His main research interests are in the area of plant physiological ecology, especially related to mineral nutrition and water relations. He was the manager of the Sarah Racine Root Research Laboratory at Tel-Aviv University.

Rosa Gálvez is an Assistant Professor at the Universidad Autónoma de Madrid (UAM). PhD in Biology in 2010. Since 2019,Entomologist specialized in the study of vector-borne diseases. Since 2013, Dr. Gálvez has initiated a line of research based on the use of arthropods as an educational resource in education. She has designed two projects, "Sponsor a queen" project related to the breeding and observation of queen ants as an innovative educational resource and "Flebocollect" citizen science project in which students themselves collaboratively manufacture phlebotomine sandfly light traps with recycled materials through a DIY "do-it-yourself" model.

Yiannis Georgiou is a member of the Special Teaching Staff at the Department of Communication & Internet Studies of the Cyprus University of Technology. He holds a bachelor's degree in *Elementary School Teaching*, a master's degree in *Science & Environmental Education* from the University of Cyprus, and a PhD in *New Technologies for Communication and Learning*. During his studies he served as Research Associate at various research groups, as follows: Learning in Science Research group, Cyprus Centre for Environmental Research and Education, Media, Cognition and Learning Research Group. From 2018 to 2020, he served as a Postdoctoral researcher at the Cyprus Interaction Lab of the Cyprus University of Technology, as well as at the Research Center for Interactive Media, Smart Systems & Emerging Technologies.

Yaela N. Golumbic is a Senior Researcher and the Head of citizen science research at The Steinhardt Museum of Natural History, at Tel-Aviv University. Her research lies at the intersection of science communication, science education, and citizen science, with an aim to make science more accessible and relevant to people's day-to-day lives. Yaela has developed, designed, and managed several citizen science projects, harnessing the power of the crowd for advancing science, leading co-creation processes, and examining its benefits and challenges for science and society.

Andreas Ch. Hadjichambis is Chair of the European Network for Environmental Citizenship (Cost Action CA16229), which includes 88 universities and research institutes from 39 countries and more than 134 academics, researchers, and scientists. He is the Scientific Director (Research Professor) of the Cyprus Centre for Environmental Research and Education (CYCERE) and a Researcher in the Cyprus University of Technology. He is a biologist (BSc) with MSc, MEd in education, and a PhD. He participated in many European research projects and is co-author of the biology books applied in Cyprus secondary education. His work has received many awards and distinctions in Cyprus and abroad.

Demetra Hadjichambi is head of Environmental Education Department of the Cyprus Centre for Environmental Research and Education (CYCERE). She is a biologist and holds a PhD in Environmental Education. She is working as a biology advisor in the Cyprus Ministry of Education and Culture. She participated in Biology Curriculum Committee in the framework of Cyprus Educational Reformation. She is a co-author of the Biology books in the Secondary Education of Cyprus. She has been a SCIENTIX Ambassador.

Sol Henson has worked as a coordinator and then director of the Sierra Streams Institute Education Program for over a decade. Before this, he worked as an environmental science researcher characterizing mercury fate and transport in the Deer Creek Watershed. He obtained a master's degree in hydrology at the University of California, Davis, and has worked on projects both locally and internationally regarding water resources and water resource management.

Yael Kali is a Professor of technology-enhanced learning at the University of Haifa, and a fellow of the International Society of the Learning Sciences (ISLS) since 2021. She is the founding director of two Israeli centers of research excellence—Learning In a NetworKed Society (LINKS, 2012–2018) and Taking Citizen Science to School (TCSS, 2017–2023). Using design-based research and design-based implementation research, Kali explores technology-enhanced learning and teaching at various contexts and age levels, from junior high school to higher education, and as part of teacher professional development programs. Her work focuses on the role of design principles for supporting learning and collaborative design, especially within networks of research-practice partnerships and in *teachers-as-designers* contexts. She has served as an Associate Editor for the journal *Instructional Science* from 2012 to 2020, and was a visiting scholar at the Centre for Research on Computer Supported Learning & Cognition, University of Sydney (2010–2011), and at the School of Education, University of Colorado Boulder (2018).

Nirit Lavie Alon is a researcher at the Faculty of Science and Technology at the Technion in Israel. Her research focuses on outdoor learning, environmental and climate change education, and citizen science. With many years of experience in environmental education at the Society for the Protection of Nature, the largest and oldest environmental organization in Israel, Dr. Lavie Alon now coordinates the organization's citizen science field and leads various biodiversity projects in collaboration with researchers. As part of the joint research team of the Technion and the University of Haifa, Dr. Lavie Alon investigates the integration of citizen science in schools through the TCSS project. She served as co-advisor (with Professor Tali Tal) for two master's degree students, both biology teachers, who developed a school curriculum that integrates outdoor citizen science. This innovative work will be presented in the chapter, providing valuable insights into the practical implementation of citizen science in education.

Marcos López-de-Felipe is a biologist with a master's degree in parasitology, specialized in medical entomology, vector-borne diseases, and vector control. He is part of the national Medical Entomology laboratory in the Instituto de Salud Carlos III and the Flebocollect citizen science group. He has taken part in multiple research projects concerning different vector Arthropoda, especially sand flies, black flies, and Culicidae mosquitoes, as well as other diseases vectors and/or urban pests such as ticks, cockroaches, pine moths, bedbugs, or rats. He has been one of the main researches in the NESCOTIGER citizen science project, which aimed to identify novel control strategies against the Asian tiger mosquito

(*Aedes albopictus*) in residential areas from a cultural control and citizen science perspective.

Ana I. Mora is currently a permanent Lecturer in the Department of Specific Sciences (Experimental Sciences Area), Faculty of Teacher Training, UAM. Degree in Biology (UAM) and PhD in Physical Anthropology (UAM, UCM, and UAH), with research in health education, guiding and organizing educational processes that influence knowledge, practices, and customs of individuals and their communities in their health and nutrition and ultimately in the concept of OneHealth. In this line, it participates in the "Flebocollect" citizen science project. In this project, students are made aware of the consequences of altering ecosystems in the emergence of diseases.

Ana I. Mora-Urda is a Graduate Student Researcher at the Center for Community and Citizen Science and an education doctoral student at the University of California, Davis. Jadda received her B.S. in Sustainable Agriculture and Food Systems from the University of California, Davis, and her M.S. in Environmental Studies from Green Mountain College. Before pursuing her Ph.D., Jadda taught for the Hawai'i Department of Education. Her current research focuses on collaborative and community-based approaches to scientific research that address local environmental challenges. Her primary interest lies in engaging youth in place-based science education research to protect local ecosystems and foster students' sense of empowerment and agency for environmental action. Jadda's research also explores several key areas: (1) How place-based learning can instill an ethic of care and stewardship for the environment; (2) The benefits of place-based learning when paired with local land stewardship projects; (3) The ethical and culturally responsive integration of Traditional Ecological Knowledge into place-based learning; and (4) The potential of place-based learning and land stewardship in mediating climate anxiety, especially among youth. Through her research, Jadda aims to contribute to the growing body of knowledge on the role of place in education.

Lavi Oren is a high school biology teacher and homeroom teacher with extensive experience in both classroom and outdoor learning science education. His passion for science education led him to pursue a master's degree at the Technion—Israel Institute of Technology, where he researched the impact of out-of-classroom instruction and citizen science projects on student learning. His thesis findings are presented in this chapter. Currently, Oren is a PhD candidate in the Faculty of Education in Science and Technology at the Technion. His research focuses on citizen science projects for middle and high school students. He also serves as the Biology Department Head at his school.

Evan F. Portier is the Education Program Manager at Sierra Streams Institute and has his B.S. in Environmental Science from the University of California,

Berkeley. He has previously worked as a researcher on several long-term ecology projects, focusing on tree seedling growth and mortality. He went on to work in informal education and has spent the last six years working at outdoor schools and public land agencies. His interests lie in engaging students in environmental issues that affect them and their communities, and how community science can help address such issues.

Alexandra I. Race is a Postdoctoral Scholar at the Center for Community and Citizen Science in the School of Education at UC Davis. She received her Ph.D. in Education from UC Santa Cruz, her M.S. in Biological Sciences and Educational Media Design from UC Irvine, and her B.A. in Integrative Biology from UC Berkeley. Alexandra's research interests include increasing equity and inclusion in science and field-based education, social and environmental justice approaches to teacher education, and critical sociocultural theory. She has published in multiple journals, including Ecology and Evolution, Integrative and Comparative Biology, and Sustainability. She has presented multiple times at the AERA and NARST conferences. She has worked on the UC Natural Reserve committee to increase inclusivity and access to natural reserves and field-based courses and research. Her current work examines the intersections of formal education and community and citizen science (CCS), how community builds around CCS projects, and how youth engage in a watershed-based CCS project.

Ornit Sagy received the BSc degree in computer science, the MBA degree in industrial engineering, and the PhD degree from the Department of Education in Technology and Science from the Technion—Israel Institute of Technology. She is a researcher in the Technologies in Education Graduate Program at the Faculty of Education, University of Haifa, and the Co-Director of the Taking Citizen Science to School (TCSS) research center. Her expertise is in designing and exploring technology-enhanced learning environments to promote an internal-value-based learning culture. In addition to her 13 years of experience as a software engineer in the Israeli and American high-tech industry, she has taught computer science in high schools, and graduate courses in technology-enhanced learning at the University of Haifa and the Weizmann Institute.

Yael Silver is a graduate student at the Technion—Israel Institute of Technology and the recipient of the Ariane De Rothchild prestigious scholarship. She is a lecturer at Tel Hai Academic College, and the co-founder of "Town Square Academia," a social initiative established to strengthen partnerships between diverse communities in the Galilee through a proactive, critical, and heterogeneous dialogue. Yael remains actively involved in the project, and her articles in this field have been published in The International Journal of Current Educational Research, Science and Technology Studies, and Public Understanding of Science. Her work was published, as well, in few book chapters.

Emma R. Schectman is an education doctoral student at the University of California, Davis. From personal experience, she knows how impactful informal science education experiences can be, which sparked her interest in informal science education. This led her to work as an outdoor educator in Maine for two years before returning to school to learn more about education. She received her M.S. in Environmental Conservation Education at New York University. After working in New York City as an educator involved with Community and Citizen Science projects, she became interested in how informal science programs can not only foster care for the environment but also help people to engage with research and findings about their local areas. Her research explores informal climate change education programs for adult learners and how these programs can help initiate action toward climate justice.

Hila Shefet Barkae is a biology teacher at Kyryat Hinuch Vetarboot Dror. During her master's at the Faculty of Science and Technology at the Technion in Israel, her research focused on inquiry learning, outdoor learning, and citizen science. As part of the joint research team of the Technion and the University of Haifa, Shefet Barkae investigates the integration of citizen science in schools through the TCSS (Taking Citizen Science to School) project. She developed a school curriculum that integrates outdoor citizen science and investigates it with her advisors Professor Tali Tal and Dr Nirit Lavie Alon. This innovative work will be presented in the chapter, providing valuable insights into the practical implementation of citizen science in education.

Ayelet Shavit is a philosopher of science and the recipient of the Bloomfield Prize, the Mary Currie EU fellowship, and the Israeli prize for Higher Education. Associate Professor Shavit is a faculty member at Tel Hai College and a research fellow at the Technion. Since 2019 she chairs Israel's committee for community-academia partnership in the HEC (Higher Education Council), since 2020 she also chairs the Israeli Philosophical Association, and between 2008 and 2021 she chaired the Philosophy Program at Tel Hai College. Ayelet studies the interplay of facts and values in ecology & evolution, and her practice-based conceptual analysis interlaces with her co-led *Town Square Academia* proactive initiative. Her publications appear in journals such as *Journal of Philosophy, Philosophy of Science, Philosophical Transactions of the Royal Society B*, and *Proceedings of the Royal Society B;* alongside her book *One for All?* (Magnes Press, 2008), and her co-edited books: *Stepping in the Same River Twice* (Yale University Press, 2017) and *Landscapes of Collectivity* (MIT Press, 2018).

Tali Tal is a science and environmental education professor and the Ruben and Tallu Rosen Academic Chair in Science at the Faculty of Education in Science and Technology of the Technion. She is the Head of the Pre-University Center of the Technion. She received her Ph.D. from the Technion, in 1998 and had completed a post-doc at the University of Michigan. Since 2000, she's been a faculty member at the Technion. Her research focuses on learning science in informal settings, inquiry-based learning, environmental education and learning with citizen science. Tali has published over 80 papers in prestigious research journals such as JRST, Science Education, Environmental Education Research, and the International Journal of Science Education, as well as numerous book chapters. Tal was the President of NARST, a worldwide organization for promoting science education through research and was the chair of the Ministry of Education Professional Committee of Environmental Sciences. She is one of the PIs of TCSS—Taking Citizen Science to School research center where she leads few citizen science projects. Together with Prof. Gail Richmond from Michigan State University, she is a co-PI on the Teaching Science Outdoors Urban Partnership NSF-DRK12 grant.

Dina Tsybulsky Ph.D., is an Associate Professor and Head of the Biology Education Research Group in the Faculty of Education in Science and Technology at the Technion Israel Institute of Technology. Her research interests include approaches to biology teaching and learning; nature of science and scientific practices; biology teacher education and professional development. Tsybulsky serves as an Associate Editor for JRST and is currently a member of the editorial boards of JSTE and DISER. She is an active member of several international and national committees, including her role as chair of the Outstanding Doctoral Research Award Committee at NARST. Tsybulsky is also one of the PIs of Taking Citizen Science to School Center, a PI of the Smart BioEd Lab, and a PI of several research projects funded by the EuroTech Universities Alliance, the Israeli Science Foundation, and the Israeli Ministry of Education.

Shulong Yan is a Postdoctoral Scholar at the Center for Community and Citizen Science, School of Education at UC Davis. She obtained her Ph.D. in Learning, Design, and Technology from Pennsylvania State University in 2021. Shulong's research interests include using design-based research, community-based, and collaborative learning approaches to develop equitable and inclusive formal and informal learning environments that foster students' multiple identities development and higher-order cognitive and socio-emotional skills with the support of technologies. Her work involves qualitative methods like interaction analysis, multimodal analysis, and design-based research. Shulong has published extensively in journals like Information and Learning Sciences, Sustainability, and Computer Science Education. She has collaborated on projects related to integrating extended reality in design and STEM education and community and citizen science programs for K-12 students.

Contributors

Nirit Lavie Alon Faculty of Education in Science and Technology, Technion Israel Institute of Technology, Haifa, Israel

Heidi L. Ballard University of California, Davis, CA, USA

Hila Shefet Barkae Faculty of Education in Science and Technology, Technion Israel Institute of Technology, Haifa, Israel

Ángela Bermejo Department of Specific Didactics, School of Education and Teacher Training, Universidad Autónoma de Madrid, Madrid, Spain

Michal Dvir Faculty of Education in Science and Technology, Technion Israel Institute of Technology, Haifa, Israel

Faculty of Education, University of Haifa, Haifa, Israel

Amram Eshel Faculty of Life Sciences, Tel Aviv University, Tel Aviv-Yafo, Israel

Marcos López de Felipe Laboratory of Medical Entomology, National Centre for Microbiology, Instituto de Salud Carlos III, Madrid, Spain

Rosa Gálvez Department of Specific Didactics, School of Education and Teacher Training, Universidad Autónoma de Madrid, Madrid, Spain

Yiannis Georgiou Department of Communication & Internet Studies, Cyprus University of Technology, Limassol, Cyprus

Yaela N. Golumbic The Steinhardt Museum of Natural History, Tel Aviv University, Tel Aviv, Israel

Bart de Groot Freudenthal Institute, Faculty of Science, Utrecht University, Utrecht, Netherlands

Demetra Hadjichambi Cyprus Centre for Environmental Research and Education (CYCERE), Lemesos, Cyprus

Cyprus Ministry of Education, Sport and Youth, Kimonos & Thoukididou, Nicosia, Cyprus

Andreas Ch. Hadjichambis Cyprus Centre for Environmental Research and Education (CYCERE), Lemesos, Cyprus

Cyprus Ministry of Education, Sport and Youth, Kimonos & Thoukididou, Nicosia, Cyprus

Sol Henson Sierra Streams Institute, Nevada City, CA, USA

Yael Kali Faculty of Education, University of Haifa, Haifa, Israel

Tarcisio Milagres Department of Microbiology, Immunology and Parasitology, Universidade Federal Do Rio Grande do Sul, Porto Alegre, Brazil

Jadda M. Miller University of California, Davis, CA, USA

Ana I. Mora-Urda Department of Specific Didactics, School of Education and Teacher Training, Universidad Autónoma de Madrid, Madrid, Spain

Yael Bar Nes Faculty of Education, University of Haifa, Haifa, Israel

Lavi Oren Faculty of Education in Science and Technology, Technion Israel Institute of Technology, Haifa, Israel

Jelle Boeve-de Pauw Freudenthal Institute, Faculty of Science, Utrecht University, Utrecht, Netherlands

Evan F. Portier Sierra Streams Institute, Nevada City, CA, USA

Alexandra I. Race University of California, Davis, CA, USA

Ornit Sagy Faculty of Education, University of Haifa, Haifa, Israel

Emma R. Schectman University of California, Davis, CA, USA

Ayelet Shavit Technion Israel Institute of Technology, Haifa, Israel

Tel-Hai College, Upper Galilee, Israel

Yael Silver Technion Israel Institute of Technology, Haifa, Israel

Tel-Hai College, Upper Galilee, Israel

Haya Ben Simon Faculty of Education in Science and Technology, Technion Israel Institute of Technology, Haifa, Israel

Tali Tal Faculty of Education in Science and Technology, Technion Israel Institute of Technology, Samuel Neaman Institute, Haifa, Israel

Dina Tsybulsky Faculty of Education in Science and Technology, Technion Israel Institute of Technology, Haifa, Israel

Shulong Yan University of California, Davis, CA, USA

Chapter 1
Introduction

Dina Tsybulsky and Tali Tal

1.1 Scope and Rationale

Citizen science is a research approach that actively involves non-scientists in authentic scientific investigations. It enables collaboration between research institutes, organizations, and the public, fostering partnerships that advance scientific knowledge. There are currently hundreds of research studies originating from various scientific institutes and universities worldwide that are being conducted in cooperation with citizen scientists. The development of citizen science contributes to our biological and ecological understanding of processes on Earth, as well as to the search for creative and effective ways to conserve nature (Cooper et al., 2007; Dillon et al., 2016; Theobald et al., 2015). The well-documented positive outcomes of citizen science for science, society, and individuals (Shirk et al., 2012; Tsybulsky, 2020) have made the field increasingly popular, particularly in ecology and environmental studies (Dickinson et al., 2012; Follett & Strezov, 2015).

In recent years, accumulating evidence has highlighted the transformative power of citizen science in education, demonstrating its potential to break down traditional barriers between science and society (Roche et al., 2020). Student participation in citizen science projects, especially when it involves more than just data collection, can transform individuals into active and informed advocates for change (Atias et al., 2023; Ballard et al., 2017). This involvement bridges the gap between research and

D. Tsybulsky
Faculty of Education in Science and Technology, Technion Israel Institute of Technology, Haifa, Israel
e-mail: dinatsy@technion.ac.il

T. Tal
Faculty of Education in Science and Technology, Technion Israel Institute of Technology, Samuel Neaman Institute, Haifa, Israel
e-mail: rtal@ed.technion.ac.il

D. Tsybulsky, T. Tal (eds.), *Citizen Science Contributions to Biology and Environmental Education*, Contributions from Biology Education Research,
https://doi.org/10.1007/978-3-031-99298-8_1

action by empowering students to protect their local environments and work towards a more sustainable future (Tal et al., 2023). Citizen science is now considered an educational practice with significant pedagogical potential. This book reflects the growing recognition of the value of school participation in citizen science as an emerging research field in the learning and instructional sciences (Kali et al., 2023).

A key issue in the field is the diversity of definitions and theoretical frameworks used to conceptualize citizen science. One widely accepted definition of citizen science describes it as research initiatives in which volunteer non-scientists collaborate with scientists to advance scientific knowledge (Hecker et al., 2018). However, scholars and disciplines interpret citizen science in various ways, and emphasize different aspects such as public engagement in scientific inquiry or community-driven knowledge production. This book embraces these diverse perspectives by presenting a range of interpretations aligned with different disciplinary backgrounds and research contexts. Recognizing these variations allows for a more nuanced understanding of how citizen science functions in both scientific research and education. Rather than enforcing a singular definition, this volume provides space for multiple theoretical and methodological standpoints, thus encouraging readers to critically engage with the evolving discourse in the field. This approach allows for a more inclusive exploration of citizen science by accommodating diverse perspectives and thus fostering a deeper understanding of its applications in both research and education.

Citizen Science Contributions to Biology and Environmental Education—Realities, Impact and Opportunities provides a comprehensive overview of the transformative potential of citizen science as a pedagogical practice. Featuring contributions from leading experts in the field, it presents evidence-based studies and practical applications that offer valuable insights into the best ways to design, implement, and evaluate citizen science initiatives effectively. This book is published as part of the Springer series on biology education research associated with the ERIDOB]European Researchers in Didactics of Biology] research community. This affiliation reflects the book's contribution to the ongoing scholarly discourse within the field that encourages international collaborations and research-based teaching practices in biology and environmental education. This book builds on these foundations to examine the role of citizen science in education from multiple perspectives.

Overall, this volume showcases how citizen science initiatives can enhance students' scientific reasoning and grasp of the nature of science, promote environmental awareness, and nurture a deeper connection between people and their surroundings. Unlike previous works, this book explores how citizen science can be systematically integrated into biology and environmental education, thus bridging theory and practice through diverse educational frameworks. It not only provides theoretical insights but also presents concrete case studies and innovative pedagogical approaches, offering a comprehensive resource for educators, researchers, and policymakers seeking to leverage citizen science for meaningful learning experiences. By involving students and community members in real-world scientific investigations, citizen science not only provides valuable data to scientific communities but also empowers the participants to become informed and active citizens.

1.2 The Structure of this Book

This book is structured to provide a comprehensive exploration of the integration of citizen science into biology and environmental education. The chapters present theoretical foundations, practical implementations, and its transformative potential in diverse educational settings. The book is divided into three sections:

1. **Foundations of Citizen Science in Biology and Environmental Education: Theoretical Perspectives and Implications.** This section lays the conceptual groundwork for understanding citizen science within biology and environmental education. It examines how citizen science has evolved, its theoretical underpinnings, and its potential as a pedagogical tool. Chapter 2 by Yaela N. Golumbic explores the mainstreaming of citizen science in classrooms by analyzing the challenges and strategies for effectively embedding citizen science within formal education. Chapter 3 by Yael Bar Nes and colleagues introduces the concept of expansive framings in citizen science, and shows how well-designed projects can cultivate student agency, critical thinking, and deeper engagement with scientific inquiry. Chapter 4 by Andreas Ch. Hadjichambis and colleagues discusses the role of citizen science in fostering environmental citizenship, and emphasizes its contribution to sustainability education, green transition policies, and participatory decision-making.
2. **Integrating Citizen Science into K-12 Biology and Environmental Education: Opportunities and Challenges.** This section focuses on the practical integration of citizen science into K-12 education. It examines how such initiatives shape students' engagement with scientific practices, reasoning, and inquiry. Chapter 5 by Michal Dvir and Dina Tsybulsky highlights the role of citizen science data in extending students' understanding of scientific frontiers, increasing data literacy, and promoting authentic scientific inquiry. Chapter 6 by Nirit Lavie Alon and co-authors presents case studies demonstrating how field-based citizen science projects enhance ecological literacy and inquiry-based learning, and offers evidence-based recommendations for educators. Chapter 7 by Haya Ben Simon and Dina Tsybulsky examines the pedagogical potential of citizen science projects in cultivating scientific reasoning, statistical thinking, and a deeper grasp of the nature of science. Chapter 8 by Bart de Groot and Jelle Boeve-de Pauw explores the critical role of teachers in citizen science implementation, by identifying the key competencies and institutional supports needed for successful integration.
3. **Fostering Engagement and Identity through Citizen and Community Science.** This section shifts the focus to broader socio-educational dimensions, and explores how citizen and community science promote engagement, scientific identity, and environmental stewardship. Chapter 9 by Jadda M. Miller and colleagues investigates the impact of citizen and community science initiatives in elementary schools. It highlights their role in shaping students' scientific identity and developing collaborative partnerships. Chapter 10 by Rosa Gálvez and her team presents *Flebocollect*, a citizen science project that engages students in

studying the spread of leishmaniosis disease, which links ecosystem alterations to global health challenges. Chapter 11 by Yael Silver and Ayelet Shavit critically examines the distinctions between community science and citizen science, while arguing for a conceptual and activist-based differentiation that has significant implications for education and research.

This structure ensures that the book offers both theoretical depth and practical guidance, equipping educators, researchers, and policymakers with insights into the transformative potential of citizen science in biology and environmental education.

1.3 Audience

This book targets a wide range of audiences from researchers and educators, to anyone interested in citizen science and its contributions to biology and environmental education. It provides readers with state-of-the-art information on the latest advances in research and practice, and optimal ways to participate in citizen science endeavors. The comprehensive coverage and diverse perspectives presented in this book make it a valuable resource for disseminating citizen science education on an international scale.

References

Atias, O., Kali, Y., Shavit, A., & Baram-Tsabari, A. (2023). Meaningful participation of schools in scientific research through contributory citizen science projects. *Science Education, 107*(5), 1163–1192. https://doi.org/10.1002/sce.21800

Ballard, H. L., Dixon, C. G. H., & Harris, E. M. (2017). Youth-focused citizen science: Examining the role of environmental science learning and agency for conservation. *Biological Conservation, 208*, 65–75. https://doi.org/10.1016/j.biocon.2016.05.024

Cooper, C. B., Dickinson, J., Phillips, T., & Bonney, R. (2007). Citizen science as a tool for conservation in residential ecosystems. *Ecology and Society, 12*(2). http://www.jstor.org/stable/26267884

Dickinson, J. L., Shirk, J., Bonter, D., Bonney, R., Crain, R. L., Martin, J., Phillips, T., & Purcell, K. (2012). The current state of citizen science as a tool for ecological research and public engagement. *Frontiers in Ecology and the Environment, 10*(6), 291–297. https://doi.org/10.1890/110236

Dillon, J., Stevenson, R. B., & Wals, A. E. (2016). Introduction to the special section moving from citizen to civic science to address wicked conservation problems. Corrected by erratum 12844. *Conservation Biology, 30*(3), 450–455. https://doi.org/10.1111/cobi.12689

Follett, R., & Strezov, V. (2015). An analysis of citizen science-based research: Usage and publication patterns. *PLoS One, 10*, e0143687. https://doi.org/10.1371/journal.pone.0143687

Hecker, S., Bonney, R., Haklay, M., Hölker, F., Hofer, H., Goebel, C., et al. (2018). Innovation in citizen science–perspectives on science-policy advances. *Citizen Science: Theory and Practice, 3*(1), 4.

Kali, Y., Sagy, O., Matuk, C., & Magnussen, R. (2023). School participation in citizen science (SPICES): Substantiating a field of research and practice. *Instructional Science, 51*(5), 687–694.

Roche, J., Bell, L., Galvão, C., Golumbic, Y. N., Kloetzer, L., Knoben, N., Laakso, M., Lorke, J., Mannion, G., Massetti, L., Mauchline, A., Pata, K., Ruck, A., Taraba, P., & Winter, S. (2020). Citizen science, education, and learning: Challenges and opportunities. *Frontiers in Sociology, 5*, 613814. https://doi.org/10.3389/fsoc.2020.613814

Shirk, J. L., Ballard, H. L., Wilderman, C. C., Phillips, T., Wiggins, A., Jordan, R., McCallie, E., Minarchek, M., Lewenstein, B. V., Krasny, M. E., & Bonney, R. (2012). Public participation in scientific research: A framework for deliberate design. *Ecology and Society, 17*(2). http://www.jstor.org/stable/26269051

Tal, T., Barkae, H. S., & Alon, N. L. (2023). From school-based citizen science to transition-driven activism with the community. In J. Diamond & S. Rosenfeld (Eds.), *Amplifying informal science learning* (pp. 141–154). Routledge.

Theobald, E. J., Ettinger, A. K., Burgess, H. K., DeBey, L. B., Schmidt, N. R., Froehlich, H. E., Wagner, C., HilleRisLambers, J., Tewksbury, J., Harsch, M. A., & Parrish, J. K. (2015). Global change and local solutions: Tapping the unrealized potential of citizen science for biodiversity research. *Biological Conservation, 181*, 236–244. https://doi.org/10.1016/j.biocon.2014.10.021

Tsybulsky, D. (2020). Self-reported reasons for participating in pro-environmental citizen science activities: A case study of butterfly monitoring in Israel. *Frontiers in Education (section STEM Education)*. https://doi.org/10.3389/feduc.2020.00116

Part I
Foundations of Citizen Science in Biology and Environmental Education: Theoretical Perspectives and Implications

Chapter 2
Mainstreaming Citizen Science in the Classroom

Yaela N. Golumbic

2.1 Introduction

Citizen science has showcased numerous advantages for promoting learning in formal education. It has been demonstrated as a multifaceted tool for enhancing learning, enriching the educational process and fostering a deeper connection between students and the subjects they are studying (Kloetzer et al., 2021; Williams et al., 2021). Benefits of incorporating citizen science in schools not only advance scientific knowledge, but also cultivate a range of skills and values that are vital for students to thrive as responsible, informed citizens (Ballard et al., 2017a; Bela et al., 2016; Groulx et al., 2017). However, while citizen science has grown exponentially in the realm of scientific research (Pelacho et al., 2021), it remains at the periphery of formal science education. This can be attributed to the numerous difficulties associated with incorporating citizen science into school settings, including logistical limitations, time constraints and the crowded curriculum (Aristeidou et al., 2023; Doyle et al., 2018; Preston et al., 2023).

Citizen science thrives within educational environments when it is orchestrated by passionate and committed educators with exceptional skills. These educators serve as the guiding forces behind successful citizen science initiatives in schools, playing a pivotal role in fostering meaningful learning experiences for their students (Jenkins, 2011; Na et al., 2022). Dedicated teachers also serve as facilitators, bridging the gap between the scientists leading the scientific research and the students participating in the initiatives whilst learning about science (Lorke et al., 2019). Yet, for many teachers, engaging in citizen science can be quite challenging. It demands significant time, effort, and, in some cases, expertise that they may not possess.

Y. N. Golumbic (✉)
The Steinhardt Museum of Natural History, Tel Aviv University, Tel Aviv, Israel
e-mail: Yaelago@tauex.tau.ac.il

D. Tsybulsky, T. Tal (eds.), *Citizen Science Contributions to Biology and Environmental Education*, Contributions from Biology Education Research,
https://doi.org/10.1007/978-3-031-99298-8_2

Science teacher training typically does not include instruction or skill building in integrating citizen science into schools, making such an effort a demanding process. Hence, choosing citizen science as an educational tool can seem unconventional when compared to more traditional teaching methods.

Nevertheless, there is no objectively apparent reason preventing the integration of citizen science into education. In fact, it frequently aligns with the educational objectives, offering not only theoretical knowledge but also tangible, experiential, hands-on learning opportunities (National Academies of Sciences Engineering and Medicine, 2018). As articulated by the Organization for Economic Co-operation and Development (OECD) Learning Framework 2030, as societies are rapidly and profoundly changing, broader education goals are needed which address environmental, economic and social challenges. The role of education is thus to equip students with the skills they need to develop agency and become active, responsible and engaged citizens (OECD, 2018). In accordance, the new Program for International Student Assessment (PISA) 2025 Science Framework, have recently added the concept of agency to be measured in future assessments. It has further added the scientific competency "research, evaluate and use scientific information for decision making and action" (OECD, 2023), which directly relates to activities undertaken during participation in citizen science projects. Ultimately, the adoption of citizen science as an educational tool reflects a shift towards more experiential, inquiry-based, and student-centered learning models. In line with educational developments in the last decade, it acknowledges that effective education is not solely about transmitting knowledge but also about fostering skills, curiosity, and a sense of agency in students (Kali et al., 2023; Shah & Martinez, 2016). In this way, citizen science offers a valuable avenue for educators to break away from traditional models and embrace innovative approaches that better prepare students for the challenges of the future.

This chapter will explore opportunities for mainstreaming citizen science in the classroom. It will discuss the benefits and challenges of this approach and building from past examples, will offer viable solutions for making citizen science more accessible and adaptable for teachers. Mainstreaming citizen science in the classroom means the establishment of citizen science as a widespread and routine educational tool across diverse schools and subjects, fostering a culture of active student participation, collaboration, and real-world scientific engagement.

2.2 Benefits of Citizen Science in the Classroom

Several studies have investigated the learning potential of participants in citizen science. These point to the many benefits for participants in promoting their scientific literacy skills and prompting their awareness and appreciation for science (National Academies of Sciences Engineering and Medicine, 2018; Phillips et al., 2018). Learning through citizen science can take place in a range of settings, including online platforms, families, communities and formal and nonformal education environments (Kloetzer et al., 2021). While similar learning outcomes have been documented in varies settings, the context of participation plays a crucial role. Contextual factors such as the nature of the project, the community involved and the environment in which it takes place, can shape and influence the effectiveness of the learning experience and the resulting outcomes (Peter et al., 2021; Phillips et al., 2019). As such, learning outcomes in the context of formal education, and even more so, within school environments may yield distinct results in comparison to learning in informal or community-based projects. Moreover, even within schools, outcomes may vary based on the resources available for students, the discussions and assignments conducted by students and teachers' instruction. For example, in the Helsinki Urban Rat Project, students collected presence-absence data on urban brown rats using track plates. As students were asked to place the track plates in places where rats are likely to move, they were compelled to reflect on the location of the study from a rats' perspective and consider how they may use the space. This consequently resulted in student increased interest, knowledge and understanding of rat behavior, ecology and habitat (Aivelo & Huovelin, 2020). In the Radon Home Survey, students engaged in a curriculum-based program where they learned about radon gas and its health dangers, conducted measurements in their homes, and analyzed the data. Learning outcomes aligned with the content studied, yet due to the practical and relevance of the study to students' lives, extended to transfer and application of scientific knowledge, stewardship, and self-efficacy outcomes (Golumbic et al., 2023; Tsapalov et al., 2020). Finally, following participation in the Hummingbird Project, Williams et al. (2021) demonstrate student's knowledge increase about hummingbird identification, diets, and migratory patterns. Yet specify, these varied at the teacher level based on their pedagogical style and project implementation.

As discussed, student learning outcomes from citizen science are diverse and inconsistent across different initiatives. While the potential outcomes cross disciplinary, conceptual, procedural and cognitive domains, research indicates that often a particular project will only demonstrate a select few (Phillips et al., 2018). The outcomes generally align with the project's topic and overarching goals, learning resources and implicit instruction towards particular learning goals. This points to the importance of pre-planning the type and scope of engagement of students, and designing learning resources that align with the intended learning goals. This idea was highlighted in a report by the National Academies of Sciences Engineering and Medicine (2018), emphasizing the need to "design for learning". Herodotou et al. (2018) adopted a design-based research (DBR) methodology for the iterative design of online citizen science tools aimed to enhance learning about the scientific method through setting up, running, managing or contributing to citizen science.

In recognition of the many advantages of citizen science within school environments and in an attempt to better organize and design for this cause, a number of initiatives have been established with the aim of facilitating and guiding learning through the integration of citizen science into schools. These include the Taking Citizen Science to Schools (TCSS)[1] center in Israel and Learning by Doing[2] initia-

[1] https://www.tcss.center/english

[2] https://lbdscience.com/

tive in Australia. Other science research centers, such as the Cornell Lab of Ornithology[3] have dedicated programs designing teacher resources and training them in citizen science practices and ways to incorporate this with students. A dedicated funding scheme "sparkling science" has been devoted for this cause in Austria, funding scientific institutions cooperation with educational institutions in developing and delivering citizen science programs. Such initiatives have yielded several teacher-scientist collaborations, designed curriculum-based programs and lesson plans, engaged hundreds of students in citizen science initiatives and expanded knowledge on learning through citizen science (Kali et al., 2020; Preston et al., 2023; Sparkling Science, 2020; Trautmann et al., 2019). However, despite commendable efforts, their ability to create a substantial impact is constrained by working with a limited number of groups and limited capacity. Ultimately, citizen science has not gained widespread adoption in schools, due to the many challenges and breaking of boundaries associated with its implementation. This topic will be elaborated upon in the next section.

2.3 Challenges in Implementing Citizen Science in the Classroom

Challenges in implementing citizen science in the classroom are many. From logistical complexities to curriculum alignment, integrating a new pedagogy challenges teachers and schools (Fig. 2.1). In this section I will discuss these challenges from three perspectives: those of teachers, those of citizen science project leaders and the overall challenge of scaling practices.

2.3.1 Teachers' Perspectives

Teachers play a pivotal role in guiding students through the learning process in the classroom and serve as gatekeepers in determining students' learning experiences and activities. They are therefore often the ones deciding if and when students will participate in citizen science initiatives, facilitating their involvement and learning (Lorke et al., 2019). To effectively fulfill this role, teachers must first cultivate confidence in their own abilities and knowledge. This allows teachers to explain complex ideas clearly, lead discussions, and assist students in making meaningful contributions to the scientific research.

Teachers perceived lack of expertise and limited scientific background, is thus the first challenge for mainstreaming citizen science in the classroom. According to the US National Survey of Science and Mathematics Education (NSSME+), limited

[3] https://www.birds.cornell.edu/k12/setup/

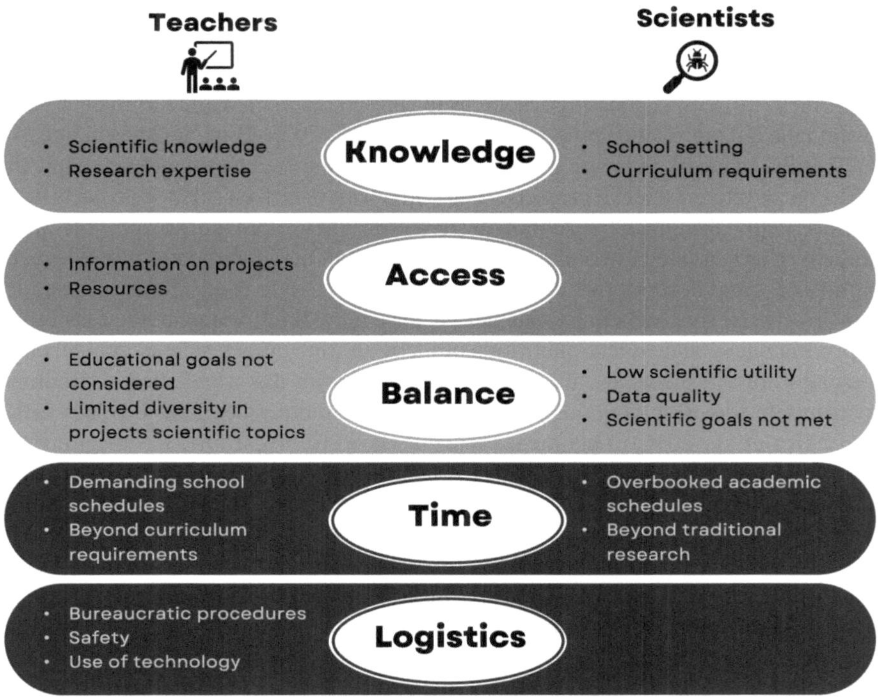

Fig. 2.1 Challenges in implementing citizen science in the classroom from the perspective of teachers and project leaders/scientists

numbers of science teachers have degrees in science or mathematics, with percentages increasing across grade levels; with under 5% for elementary school, 42% for middle school and 79% for high schools (Banilower et al., 2018). Given teachers limited accreditation in scientific disciplines, they may find it difficult to understand the intricacies of certain scientific citizen science projects, particularly in content beyond the standard curriculum. This could influence their ability to support students in participating in such projects, answer their students' questions with authority and thus, may serve as a barrier to their engagement in citizen science (Preston et al., 2023). Even those who do have a scientific background, may find a particular task quite demanding (e.g. identify an unknown variety of organisms) and have concerns about their ability to support their students in such tasks (He & Wiggins, 2017). To resolve this gap, some citizen science projects have designed workshops, training session and professional development opportunities, guiding teachers on the methodologies and scientific background of the projects' specific field (e.g. Kelemen-Finan et al., 2018; Schuttler et al., 2019). While these are beneficial in enhancing teachers' understanding of the scientific concepts and methodologies involved, they are often not scalable or sustainable over time, due to resource and time constraints and the limited reach of such programs.

An additional challenge raised by teachers is the inaccessibility of information and resources for citizen science. Techers note difficulties in finding accessible information about suitable citizen science projects to engage their students with, in addition to difficulties locating resources to support teachers with learning activities around the citizen science projects (Aristeidou et al., 2023; Braz Sousa et al., 2024).

Thirdly, teachers point to a mismatch between their educational needs and the offerings of citizen science projects. As the majority of citizen science projects were not originally designed for school involvement, they often do not provide well-defined criteria related to volunteers' (including students) qualifications, the level of complexity, and descriptions of learning outcomes (Doyle et al., 2018). These factors are crucial for teachers to consider because they directly impact the feasibility, appropriateness, and educational value of incorporating citizen science projects into the classroom. Projects are also often not aligned with the established curriculum, focusing on scientific domains that are beyond the requirements of the class syllabus (Silva et al., 2016). This means teachers need to invest time in designing the integration of citizen science projects to ensure they meet educational requirements without sacrificing their scientific authenticity. Alternatively, teachers may choose to engage with such projects as an elective, or an extra curriculum activity, contributing further to the burden and time constraints noted above. An additional mismatch noted by teachers is that of the limited scientific scope of citizen science projects. Whilst the science curriculum usually incorporates a diversity of scientific disciplines (physics, chemistry, biology, geology etc.), a limited number of projects are available in a few of these topics. In fact, research has indicated that 75–90% of citizen science projects to date have a clear focus on the life sciences (De-Groot et al., 2022; Hecker et al., 2018; Pelacho et al., 2021). While these projects offer valuable opportunities for hands-on learning and engagement in topics such as ecology, biology, and environmental science, the limited diversity of project topics inadvertently restricts the opportunities for teachers to engage their students in citizen science. Even within the natural sciences, greater curricular emphasis is put recently on contemporary scientific practices such as molecular biology, genetics, and biotechnology (Bybee, 2014; National Research Council, 2013) rather than natural history, biodiversity or ecology which dominate the citizen science field. This further restricts the alignment of citizen science with the school curriculum.

An additional challenge teachers struggle with, for integrating citizen science in the classroom, relates to time constraints. This challenge arises because teachers have demanding schedules that encompass lesson planning, classroom instruction, assessments, and administrative responsibilities. Integrating citizen science into their already crowded curriculum while also ensuring alignment with educational standards and preparing for national exams (especially in high schools) can be challenging. Such integration also necessitates allocating additional time for planning, running activities, and facilitating student engagement. Time teachers often feel they do not have to spare (Aristeidou et al., 2023; Silva et al., 2016). Given these time pressures, educators may perceive the incorporation of citizen science as an additional burden, making it challenging to prioritize and implement effectively.

The final challenge for integrating citizen science into schools from teachers' perspective relates to logistics, with bureaucratic procedures like obtaining parental and school permission and use of technology posing significant hurdles. The successful integration of citizen science is contingent upon the administration's support for new teaching strategies, an approval teachers seemed uncertain if their school would support (Braz Sousa et al., 2024; He & Wiggins, 2017). The use of technology was indicated as a challenge with concerns about ensuring student online safety, access to technology and concerns about letting students use personal cell phones which is inconsistent with the usual schools' policies (He & Wiggins, 2017; Preston et al., 2023). Student physical safety in addition to gaining permission to venture beyond school grounds and the associated paperwork, serves as an additional challenge (Aristeidou et al., 2023; Preston et al., 2023). Lastly, the availability of financial support (such as funds to cover the costs of transporting students to observation sites), as well as the provision of necessary equipment, tools, and facilities, further complicate the engagement of teachers in some citizen science projects.

Ultimately, in order for the citizen science to be integrated in schools, benefits have to outweigh the challenges. This is sometimes not the case, as demonstrated by (Aristeidou et al., 2023) who interviewed teachers engaged in citizen science projects with students. Teachers acknowledged difficulties in sustaining their own motivation and engaging other educators in citizen science activities, primarily due to the absence of incentives and support, including assistance with activity design, funding, and collaborative opportunities.

2.3.2 Citizen Science Project Leaders' Perspectives

Running citizen science projects in schools presents a unique set of challenges for scientists and project leaders,[4] as well. While some of these challenges may overlap with those faced by teachers, certain issues are uniquely tied to the scientist's perspectives and professional development.

Firstly, citizen science project leaders may find themselves lacking the necessary resources and knowledge to seamlessly integrate projects into school settings. Scientist's unfamiliarity with the educational intricacies, lesson design and classroom dynamic can make it difficult to deliver effective and engaging citizen science projects within schools. Scientists may also struggle to communicate complex scientific concepts in a manner that resonates with students and aligns with established pedagogical methods (Brownell et al., 2013). Furthermore, unlike traditional research environments, schools have distinct operational dynamics, ranging from diverse student abilities to varied classroom resources. This discrepancy often demands adaptability on the part of scientists, who may need to tailor their projects

[4] In this section the terms "scientist" and "project leader" are used interchangeably. While acknowledging not all project leaders are in fact scientists, for the purpose of the arguments presented here, they are considered to share key qualities.

to fit the specific constraints and capacities of school environments (Preston et al., 2023). As discussed by Dittmann et al. (2023), projects need to consider the realities in which schools operate, these includes teaching schedules, school curricula, vacations, etc. In order to do so, project leaders have indicated the need for guidance and clarity on teachers' expectations, perspectives and availabilities, in addition to students' learning needs, emphasizing the importance of these, for effective project adaptation (Preston et al., 2023).

The second challenge many project leaders encounter is balancing the scientific and educational aspects of citizen science programs. Whilst many scientists express motivations linked to fostering students' better understanding and connection to science (Atias et al., 2023a), it's crucial to note that their primary objective remains the advancement of scientific knowledge. However, achieving this dual goal isn't always straightforward. The inherent tension lies in balancing the educational benefits for students with the scientific rigor and progress the project aims to contribute to (Bopardikar et al., 2023; Roche et al., 2020). While the primary goal is to engage students in authentic scientific research, the academic requirements and constraints within schools may necessitate adjustments to the original project design. Scientists must navigate these compromises skillfully, ensuring that the essence of the research is retained while aligning with educational objectives. Unfortunately, in some case scientists report low utility of students' engagement with citizen science for the scientific advancement of their projects due to insufficient data collection, low data quality or inefficient allocation of time and resources (Jadallah & Wise, 2023; Preston et al., 2023; Zoellick et al., 2012). Scientists have also expressed concerns about the absence of opportunities for career development, whilst engaging in citizen science projects in schools, noting that the research questions in some instances are perceived as less meaningful, posing an additional challenge to their professional growth (Atias et al., 2023a).

Thirdly, similar to teacher challenges, time constraints inherent in the academic calendar in addition to the overbooked schedules of many scientists and project leaders, may limit the duration and frequency of their involvement (Esch et al., 2020). This is particularly true for ongoing programs, which require several get-togethers and interactions between project leaders, educators and students, or other demanding tasks, such as timely analysis of many samples (Dittmann et al., 2023). Time is also a constraint when many adaptations and adjustments to the original project design are required in order to engage students.

In essence, scientists leading citizen science initiatives in schools must navigate these challenges with a nuanced understanding of the educational landscape. This involves not only adapting their projects to suit the educational context but also addressing resource limitations and time constraints to ensure a meaningful and sustainable collaboration between the scientific and educational aspects of citizen science in school settings. It is therefore understandable why many scientists perceive engagement with citizen science in schools as a multifaceted challenge and may choose not to engage in the process. While scientists often aspire to foster a genuine connection to science among students, the reality of these complexities sometimes underscores the good intentions and may cause scientists to withhold such engagement.

2.3.3 *The Scaling Challenge*

As outlined above, the challenges for integrating citizen science in the classroom are numerous both from the perspectives of scientists and those of teachers. On an individual level, solutions may be devised to tackle these challenges within specific contexts. Indeed, we witness many successful implementations of citizen science in schools (e.g. Saunders et al., 2018; Schuttler et al., 2019; Spicer et al., 2020), and many effective collaborations between teachers and scientists (e.g. Atias et al., 2023b; Kermish-Allen et al., 2019). Yet, these implementations are often time and resource intensive and hence, do not serve as scalable solutions.

Finding strategies that cater to all teachers is complex, given the variations in their levels of knowledge, experience, and commitment. As the educational landscape varies across institutions, regions, and curricular frameworks, the needs of teachers also change considerably. Research has shown that often teachers who choose to engage with citizen science in the classroom are experienced educators with at least 10 years of teaching experience, many of which serve as subject coordinators or teacher trainers (Golumbic et al., 2023; Koomen et al., 2014). Likewise, scientists have many needs and a varied experience in engaging with young people. Often those who choose to engage with citizen science, do so with an educational goal in mind, sometimes relinquishing their own scientific objectives (Atias et al., 2023a; L'Astorina et al., 2023). Ultimately, individuals who choose to participate in citizen science projects in schools (both teachers and scientists), stand out for their unique qualities, demonstrating not only high motivation but also a visionary mindset. These individuals often bring a special set of characteristics that drive their engagement. Regrettably, such qualities are not universally shared among all teachers and scientists. Achieving widespread integration of citizen science in classrooms is therefore challenging—transforming the exceptional into a routine process. To address this challenge a comprehensive, adaptable and straightforward approach is necessary. One that considers the diverse settings and requirements of different schools and science stakeholders. The subsequent section will provide a detailed explanation of the process I suggest in order to achieve this.

2.4 Mainstreaming CS in the Classroom

When considering mainstreaming citizen science in the classroom, the challenges discussed above must be addressed, ensuring scalable implementation of projects across diverse educational settings. Drawing upon past insights, existing literature, and both successful and unsuccessful experiences, three solutions emerge: moderate, co-create and simplify (Fig. 2.2).

While each of these solutions can be implemented independently, their effectiveness is maximized when applied jointly. First, the introduction of a moderation or intermediary layer to provide essential support and guidance; second, the encouragement of collaboration and co-creation between scientists and teachers to tailor

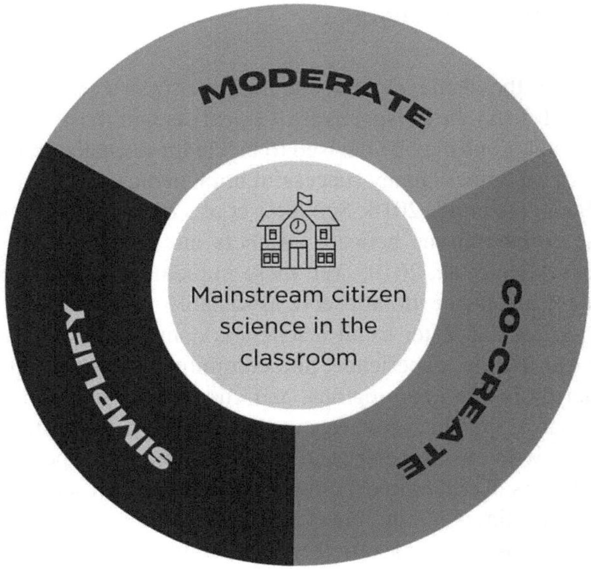

Fig. 2.2 Three solutions for mainstreaming citizen science in the classroom. Their collective implementation is suggested to yield a comprehensive approach for implementation of citizen science across diverse educational settings

projects to educational goals; and third, increased accessibility and options for a limited implementation approach aimed to make project application simple and effortless within varied curricular constraints. This multifaceted approach seeks to enhance the accessibility and effectiveness of citizen science initiatives, fostering a more inclusive and widespread integration within school environments.

2.4.1 Moderation

As discussed above, some of the main challenges expressed by teachers and scientists alike, are the lack of time and knowledge for adjusting and modifying citizen science projects to the changing needs of schools. These challenges can be addressed by introducing an intermediary layer such as an organization or facilitator, who can serve as a moderator between teachers and scientists who lead citizen science projects (Zoellick et al., 2012). Such intermediaries can provide essential support by offering guidance, resources, and assistance in project implementation. They can address the varying levels of expertise and commitment among teachers, assist with the adjustment of projects to educational needs and streamline the implementation of projects to ensure the experience is enriching for students and data usable for scientists.

The concept of facilitators in citizen science has been first discussed by Lorke (2019) who defined facilitators as: "*People who train or educate participants in a citizen science project, or lead groups of participants. Can include scientists, teachers, nature guides, museum educators etc.*" This concept was elaborated from individuals to groups or external organizations responsible for recruiting, training, and managing citizen science volunteers, facilitating interactions between scientists and participants and providing support for projects (Lin Hunter et al., 2023; Salmon et al., 2021). The structure provided by facilitating groups has been invaluable to project implementation, as they bring additional expertise and skillsets such as management and technical services, increase community participation and serve as a bridge between scientists and the community (Salmon et al., 2021). In the context of education, Smith et al. (2023) report on an array of cognitive, affective, and behavioral outcomes supported by the partnership of two facilitating organizations, Girl Scouts of the USA and SciStarter citizen science platform. They point to the potential of facilitator organizations to leverage existing citizen science projects, to promote learning goals, foster civic science education and inspire community action.

Organizations who often take leading roles in delivering and facilitating citizen science projects, include natural history museums, zoos, botanical gardens, and similar science and nature public engagement organizations (Ballard et al., 2017b; Chen & Sun, 2018; Sforzi et al., 2018). This is due to their vast experience in science communication and collaboration with amateur-expert naturalists, alongside their commitment to science, education and public engagement. Such organizations often have teams focusing on both scientific and public-facing goals which make them ideal to lead or serve as moderators in citizen science projects. In this dual role, they effectively bridge the gap between scientific research, public engagement and education, contributing to the multifaceted goals of citizen science initiatives (Sforzi et al., 2018). The introduction of a moderator layer to the implementation of citizen science projects, serves as a promising avenue to alleviate the challenges posed by teachers and scientists as to their capacity and knowledge for running citizen science projects in schools. The moderator can streamline the process, provide support and guidance to teachers and scientists involved and ensure the initiatives remain adaptable, relevant, and effectively tailored to the dynamic educational needs of schools. Yet, challenges in the moderator role may arise, related to communication and coordination between stakeholders. This points to the need to coordinate roles and expectations and maintain active engagement of teachers and scientists alongside the moderator. The TCSS center (introduced above), implemented an approach employing science education researchers as moderators of citizen science projects (Benichou et al., 2022). In this role, moderators promoted active and ongoing involvement of teachers and scientists in setting and reaching project goals and facilitated the development of partnerships and collaborations, a theme elaborated upon in the next section.

2.4.2 Collaboration and Co-Creation

In response to the challenge posed by the misalignment between the educational needs of teachers and the offerings of citizen science projects, as well as the need to strike a balance between the scientific and educational outcomes of such programs, a collaborative and co-creation approach can be adopted. This involves fostering a partnership between teachers and scientists in the design and implementation of citizen science initiatives and resources (Roche et al., 2020). Through co-creation, educators can actively contribute their pedagogical expertise, ensuring that the projects align with educational goals and curriculum standards, and cater to the diverse needs of students. Scientists contribute their scientific knowledge to enrich the learning experience and to design robust scientific protocols.

Teacher-scientist partnerships similarly known as school-university partnership or research-practice partnerships, have a long tradition in science education as a means of bridging the gap between institutions and their practices. Types of partnerships include placement of university staff within school settings, teacher professional development in academic institutions, and co-creation of tools, lessons and materials (Sarmiento-Márquez et al., 2023). In some cases, partnerships include several of the above, such as the scientist-teacher partnership reported by Brown et al. (2014), in which high school science teachers participated in multipurpose professional development program. Teachers were presented with current research findings, participated in an authentic laboratory sequence and then brainstormed together with university scientists about ways to incorporate the scientific content and methods into classroom applications. Proposals were reviewed and revised by peers and other university partners and materials were made available via the project web site.

In the context of citizen science, Bopardikar et al. (2023) explain the role of Student-Teacher-Scientist Partnerships (STSPs) as one of facilitating boundary crossing. That is, establishing actions or interactions across practices which traditionally have different norms, goals and tools, in our case, the domains of scientific research and science education. When successful, such partnerships ensure students and teachers contribute to research projects whilst making science more accessible, and fostering students' knowledge and interest in science (Zoellick et al., 2012). Such partnerships have yielded curriculum support such as customized learning materials, lesson plans for data collection covering curriculum requirements and inclusion of local examples and discussions as part of the in-class tasks (Bonney et al., 2009; Harlin et al., 2019). In addition, partnerships have progressed instructional support such as videos and guides, structured data sheets and protocols to aid in data collection and reporting, and student interaction with scientists (Bopardikar et al., 2023). Tal et al. (2023) report on a co-created citizen science project developed jointly by a science teacher, middle-school students and three scientists from a university, NGO and governmental organization. Their study demonstrates the collaborative strength in initiating actions for nature conservation alongside meeting requirements of science inquiry. Ultimately, collaborative efforts enhance the overall quality and relevance of citizen science programs in educational settings, ensuring all stakeholders' needs and goals are met.

While the collaboration and co-creation approach suggested is time and labor consuming, the essence of these collaborations is to extend assistance to others. The rigorous efforts of few, are intended to amplify the implementation of programs for those unwilling or unable to engage in an in-depth process, such as co-creation of a new project or program. In other words, dedicated teachers who are passionate and committed to citizen science and have the capacity to engage in a co-created process, contribute to the creation of curriculum aligned programs and resources, which can then be used by teachers and schools at large scale. This is crucial for addressing the other challenges discussed herein, such as time constraints and knowledge barriers, and will explored further in the following section.

2.4.3 *Simplification of Project Implementation*

The last suggestion is to simplify the process of citizen science participation by limiting the scope of citizen science implementation in the classroom. This might stand in contrary to traditional educational beliefs that advocate for longitude processes to promote meaningful learning. Yet it is proposed to address the challenges presented by teachers as to accessibility and logistic challenges, in addition to the time constraints raised by both scientists and teachers. Simplifying the engagement in citizen science includes providing clear and detailed information to teachers about the project, its purpose, timeframe and requirements, the support provided, age level it is appropriate for and relevance to students' learning and local context (Doyle et al., 2018). It also means helping teachers access resources, embed them seamlessly into the classroom environment, negating the need for special preparations. Students then participate in predefined activities integrated into the standard curriculum, requiring only a limited number of short-term lessons. This approach is designed to minimize the burden on teachers by reducing preparations and eliminating the need for special permissions or direct communication with scientists and other external organizations. For scientists this means that following the initial investment in designing engaging projects and clear protocols, their ongoing effort can be reduced. By establishing a well-structured and self-sustaining framework for engaging teachers and students, the continued progress of the project is enabled without requiring constant, direct intervention.

Examples for simplifying and making citizen science accessible for teachers are resources created by the Learning by Doing initiative who identified citizen science projects suited for schools and mapped them to the Australian Syllabus across their four scientific themes: Physical World, Earth and Space, Living World and Chemical World (Learning By Doing, 2021b; Preston et al., 2023). A more detailed resource focuses on one particular project, Big City Bird[5], and includes information about project aims, main curricular links, ways to engage students and an overview on use of the projects' app (Learning By Doing, 2021a). Additional educator resources are available from the Cornell Lab K-12 Education team, offering nature-based inquiry curricula, science lesson plans, student guidebooks and more, aimed to build science skills while inspiring young people to connect to local habitats, explore biodiversity, and engage in citizen-science projects (Cornell Lab K-12 Education, 2024).

[5] https://www.spotteron.com/bigcitybirds

While further investigation is needed to assess the efficacy of these resources, the streamlined process they offer ensures that everything is pre-made and ready for integration into the classroom. Students can actively participate in citizen science experiences bearing its fruit and benefits, without disrupting the regular flow of their educational routine. Such resources also offer an opportunity to engage in several citizen science projects throughout the course of a year or semester, whilst matching each encounter to the content being taught.

The process of simplification not only eases the access to citizen science resources and reduces the time needed for its implementation, but also plays a role in addressing logistic challenges. Limiting the scope of citizen science participation can result in smoother coordination and reduced resource requirements and administrative obligations. The simplified framework facilitates the seamless continuation of students' learning, fostering engagement and exposure to real-world scientific concepts within the bounds of school day-to-day constraints and busy academic schedules.

It is recognized that this approach entails incorporating citizen science into a less intensive educational approach, which may have associated drawbacks. It is also important to note that the suggestion made here for simplifying citizen science engagement in schools, does not diminish the effectiveness of an in-depth approach. Various examples have been discussed above highlighting the utility of comprehensive widespread student engagement in achieving diverse educational goals (Schuttler et al., 2019; Spicer et al., 2020). However, acknowledging the need for scalability in citizen science initiatives, balancing the advantages of both approaches becomes crucial in creating a citizen science framework that is not only effective but also adaptable and scalable to different educational settings and audiences.

One might inquire about the responsibility for the process of simplification of citizen science engagement, particularly in the absence of the most demanding resource—time. The resolution lies in the solutions discussed in the preceding sections—namely, moderation and co-creation. It is through the strategic implementation of these facets that simplification can be attained. By distributing responsibilities and expertise through moderation and co-creation, collaborative efforts between teachers and scientists can be managed and enhanced ensuring a shared understanding of objectives, enabling the development of simplified materials that cater to the needs of diverse educators and learners. This strategic approach not only eases the burden on individual teachers but also enhances the overall effectiveness and accessibility of citizen science initiatives in educational contexts. This approach allows for the capability of a citizen science project to be easily expanded, adapted, and implemented on a larger scale without compromising its effectiveness or quality.

2.5 Summary

Citizen science aligns well with several educational objectives, offering experiential, hands-on learning opportunities and fostering agency and a deeper connection between students and scientific concepts. Yet, the many challenges related curriculum adaptation, time constraints, accessibility and logistics prevent its mainstream acceptance. In this chapter, I attempt to address these challenges, offering a nuanced strategy for making citizen science more accessible and adaptable for teachers. The proposal for mainstreaming citizen science in the classroom builds on three existing solutions: moderation, co-creation, and simplification, integrating them into a scalable approach. I offer this as an approach to ensures the benefits and outcomes of citizen science can be extended to a broader audience or applied across various settings without a proportional increase in complexity, resource requirements, or challenges. A moderation layer provides essential support in understanding and mitigating the educational and scientific settings, collaboration and co-creation tailor projects to meet both educational and scientific goals, and simplification enhances accessibility within diverse curricular and logistical constraints.

While each of these solutions can (and indeed have) be implemented independently, their strength lies in their collective application. Moderation of a project implementation by facilitator organization, sets the stage for collaborative efforts and shared understanding between teachers and scientists. It ensures that projects align with educational and scientific goals and encourages co-creation between scientists and teachers. Through collaboration and co-creation, projects are tailored to the needs of stakeholders, making them relevant, engaging, and aligning them with curriculum requirements. This can lead to greater accessibility of projects and simplification of their implementation, for teachers who do not have the time, knowledge or commitment to engage in an in-depth process of co-creation. In a continuous cycle, the moderation can then evolve into a role of overseeing the distribution, management, and ongoing activity of the simplified citizen science projects. This may ensure consistent implementation, adaptability, and efficient functioning over time, as well as a body to which teachers can turn to for support in their classroom implementation. Overall, this approach has the potential to enhance the effectiveness and accessibility of citizen science initiatives in schools, eases the burden on individual teachers, and helps streamline citizen science in diverse educational landscapes.

References

Aivelo, T., & Huovelin, S. (2020). Combining formal education and citizen science: A case study on students' perceptions of learning and interest in an urban rat project. *Environmental Education Research, 26*(3), 324–340. https://doi.org/10.1080/13504622.2020.1727860

Aristeidou, M., Lorke, J., & Ismail, N. (2023). Citizen science: Schoolteachers' motivation, experiences, and recommendations. *International Journal of Science and Mathematics Education, 21*(7), 2067–2093. https://doi.org/10.1007/s10763-022-10340-z

Atias, O., Baram-Tsabari, A., Kali, Y., & Shavit, A. (2023a). In pursuit of mutual benefits in school-based citizen science: Who wins what in a win-win situation? *Instructional Science, 51*(5). https://doi.org/10.1007/s11251-022-09608-2. Springer Netherlands.

Atias, O., Kali, Y., Shavit, A., & Baram-Tsabari, A. (2023b). Meaningful participation of schools in scientific research through contributory citizen science projects. *Science Education, 107*(5), 1163–1192. https://doi.org/10.1002/sce.21800

Ballard, H. L., Dixon, C. G. H., & Harris, E. M. (2017a). Youth-focused citizen science: Examining the role of environmental science learning and agency for conservation. *Biological Conservation, 208*, 65–75. https://doi.org/10.1016/j.biocon.2016.05.024

Ballard, H. L., Robinson, L. D., Young, A. N., Pauly, G. B., Higgins, L. M., Johnson, R. F., & Tweddle, J. C. (2017b). Contributions to conservation outcomes by natural history museum-led citizen science: Examining evidence and next steps. *Biological Conservation, 208*, 87–97. https://doi.org/10.1016/j.biocon.2016.08.040

Banilower, E. R., Smith, P. S., Malzahn, K. A., Plumley, C. L., Gordon, E. M., & Hayes, M. L. (2018). Report of the 2018 NSSME+. https://horizon-research.com/NSSME/2018-nssme/research-products/reports/technical-report

Bela, G., Peltola, T., Young, J. C., Balázs, B., Arpin, I., Pataki, G., Hauck, J., Kelemen, E., Kopperoinen, L., Van Herzele, A., Keune, H., Hecker, S., Suškevičs, M., Roy, H. E., Itkonen, P., Külvik, M., László, M., Basnou, C., Pino, J., & Bonn, A. (2016). Learning and the transformative potential of citizen science. *Conservation Biology, 30*(5), 990–999. https://doi.org/10.1111/cobi.12762

Benichou, M., Kali, Y., & Hod, Y. (2022). Teachers' expansive framing in school-based citizen science partnerships. In *Teacher learning in changing contexts* (pp. 256–276). Routledge.

Bonney, R., Cooper, C. B., Dickinson, J., Kelling, S., Phillips, T., Rosenberg, K. V., & Shirk, J. (2009). Citizen Science: A developing tool for expanding science knowledge and scientific literacy. *Bioscience, 59*(11), 977–984. https://doi.org/10.1525/bio.2009.59.11.9

Bopardikar, A., Bernstein, D., & McKenney, S. (2023). Boundary crossing in student-teacher-scientist-partnerships: Designer considerations and methods to integrate citizen science with school science. *Instructional Science, 51*(5). https://doi.org/10.1007/s11251-022-09615-3

Braz Sousa, L., Kenneally, C., Golumbic, Y., Martin, J. M., Preston, C., Rutledge, P., & Motion, A. (2024). Teacher experiences and understanding of citizen science in Australian classrooms. *PLoS ONE, 19*(11), e0312680. https://doi.org/10.1371/journal.pone.0312680

Brown, J. C., Bokor, J. R., Crippen, K. J., & Koroly, M. J. (2014). Translating current science into materials for high school via a scientist-teacher partnership. *Journal of Science Teacher Education, 25*(3), 239–262. https://doi.org/10.1007/s10972-013-9371-y

Brownell, S. E., Price, J. V., & Steinman, L. (2013). Science communication to the general public: Why we need to teach undergraduate and graduate students this skill as part of their formal scientific training. *Journal of Undergraduate Neuroscience Education, 12*(1), E6–E10.

Bybee, R. W. (2014). NGSS and the next generation of science teachers. *Journal of Science Teacher Education, 25*(2), 211–221. https://doi.org/10.1007/s10972-014-9381-4

Chen, G., & Sun, W. (2018). The role of botanical gardens in scientific research, conservation, and citizen science. *Plant Diversity, 40*(4), 181–188. https://doi.org/10.1016/j.pld.2018.07.006

Cornell Lab K-12 Education. (2024). *Lessons & activities*. https://www.birds.cornell.edu/k12/get-started/

De-Groot, R., Golumbic, Y. N., Martínez Martínez, F., Hoppe, H. U., & Reynolds, S. (2022). Developing a framework for investigating citizen science through a combination of web analytics and social science methods—The CS Track perspective. *Frontiers in Research Metrics and Analytics, 7*. https://doi.org/10.3389/frma.2022.988544

Dittmann, S., Kiessling, T., Mederake, L., Hinzmann, M., Knoblauch, D., Böhm-Beck, M., Knickmeier, K., & Thiel, M. (2023). Sharing communication insights of the citizen science program Plastic Pirates—Best practices from 7 years of engaging schoolchildren and teachers in plastic pollution research. *Frontiers in Environmental Science, 11*, 1233103. https://doi.org/10.3389/FENVS.2023.1233103/BIBTEX

Doyle, C., Li, Y., Luczak-Roesch, M., Anderson, D., Glasson, B., Boucher, M., Brieseman, C., Christenson, D., & Coton, M. (2018). What is online citizen science anyway? An educational perspective. http://arxiv.org/abs/1805.00441

Esch, R. K., Burbacher, E., Dodrill, E., DeVanna Fussell, K., Magdich, M., Toledo, M., & Robert Midden, O. W. (2020). *Citizen science in schools: Scientists' perspectives on promise and pitfalls*. Horizon Research Inc. https://www.horizon-research.com/%0Acitizen-science-in-schools-scientists-perspectives-on-promise-and-pitfalls

Golumbic, Y. N., Peri, A., Shpak, M., Tsapalov, A., Kovler, K., Ben-Zvi, D., & Baram-Tsabari, A. (2023). Citizen science and public involvement in research combining science and society: The case of the Radon home survey. *Israeli Sociology, 24*(1).

Groulx, M., Brisbois, M. C., Lemieux, C. J., Winegardner, A., & Fishback, L. (2017). A role for nature-based citizen science in promoting individual and collective climate change action? A systematic review of learning outcomes. *Science Communication, 39*(1), 45–76. https://doi.org/10.1177/1075547016688324

Harlin, J., Kloetzer, L., Patton, D., Leonhard, C., & Leysin American School High School Students. (2019). Turning students into citizen scientists. In S. Hecker, M. Haklay, A. Bowser, Z. Makuch, J. Vogel, & A. Bonn (Eds.), *Citizen science: Innovation in open science, society and policy*. UCL Press.

He, Y., & Wiggins, A. (2017). Implementing an environmental citizen science project: Implementing an environmental citizen science project: Strategies and concerns from educators' perspectives strategies and concerns from educators' perspectives. *International Journal of Environmental and Science Education, 12*(6), 1459–1481. https://digitalcommons.unomaha.edu/isqafacpub/81

Hecker, S., Garbe, L., & Bonn, A. (2018). The European citizen science landscape – A snapshot. In S. Hecker, M. Haklay, A. Bowser, Z. Makuch, J. Vogel, & A. Bonn (Eds.), *Citizen science: Innovation in open science, society and policy* (pp. 190–200). UCL Press. https://doi.org/10.2307/j.ctv550cf2.20

Herodotou, C., Aristeidou, M., Sharples, M., & Scanlon, E. (2018). Designing citizen science tools for learning: Lessons learnt from the iterative development of nQuire. *Research and Practice in Technology Enhanced Learning, 13*(4). https://doi.org/10.1186/s41039-018-0072-1

Jadallah, C. C., & Wise, A. L. (2023). Enduring tensions between scientific outputs and science learning in citizen science. *Biological Conservation, 284*(January), 110141. https://doi.org/10.1016/j.biocon.2023.110141

Jenkins, L. L. (2011). Using citizen science beyond teaching science content: A strategy for making science relevant to students' lives. *Cultural Studies of Science Education, 6*(2), 501–508. https://doi.org/10.1007/s11422-010-9304-4

Kali, Y., Sagy, O., Lavie Alon, N., Ronit, D., & TCSS. (2020). From a Network of Research-Practice Partnerships to a multi-expertise learning and design community. In Gresalfi & Horn (Eds.), *The interdisciplinarity of the learning sciences, 14th International Conference of the Learning Sciences (ICLS)*. https://doi.org/10.13140/RG.2.2.22169.24163

Kali, Y., Sagy, O., Matuk, C., & Magnussen, R. (2023). School participation in citizen science (SPICES): Substantiating a field of research and practice. *Instructional Science, 51*(5), 687–694. https://doi.org/10.1007/s11251-023-09638-4

Kelemen-Finan, J., Scheuch, M., & Winter, S. (2018). Contributions from citizen science to science education: An examination of a biodiversity citizen science project with schools in Central Europe. *International Journal of Science Education, 40*(17), 2078–2098. https://doi.org/10.1080/09500693.2018.1520405

Kermish-Allen, R., Peterman, K., & Bevc, C. (2019). The utility of citizen science projects in K-5 schools: Measures of community engagement and student impacts. *Cultural Studies of Science Education, 14*(3), 627–641. https://doi.org/10.1007/S11422-017-9830-4/TABLES/2

Kloetzer, L., Lorke, J., Roche, J., Golumbic, Y. N., Winter, S., & Jõgeva, A. (2021). Learning in citizen science. In K. Vohland, A. Land, L. Ceccaroni, R. Lemmens, J. Perelló, M. Ponti, R. Samson, & K. Wagenknecht (Eds.), *The science of citizen science* (pp. 283–308). Springer International Publishing.

Koomen, M. H., Blair, R., Young-Isebrand, E., & Oberhauser, K. S. (2014). Science professional development with teachers: Nurturing the scientist within. *Electronic Journal of Science Education, 18*(6), 1–21.

L'Astorina, A., Davis, C., Pugnetti, A., Campanaro, A., Oggioni, A., & Bergami, C. (2023). Scientists' attitudes about citizen science at Long-Term Ecological Research (LTER) sites. *Frontiers in Environmental Science*, 1–12. https://doi.org/10.3389/fenvs.2023.1130022

Learning By Doing. (2021a). *Big city bird citizen science project mapped to the NSW stage 4 curriculum*. https://doi.org/10.5281/ZENODO.5784702

Learning By Doing. (2021b). *The NSW syllabus and citizen science – Stage 4 curriculum mapping*. https://doi.org/10.5281/ZENODO.5784688

Lin Hunter, D., Johnson, V., & Cooper, C. (2023). Diversifying large-scale participatory science: The efficacy of engagement through facilitator organizations. *Citizen Science: Theory and Practice, 8*(1), 1–14. https://doi.org/10.5334/cstp.627

Lorke, J., Golumbic, Y. N., Ramjan, C., & Atias, O. (2019). Training needs and recommendations for citizen science participants, facilitators and designers. *COST action 15212 report*. http://hdl.handle.net/10141/622589

Na, S. H., Kim, J. U., Ga, S. H., Park, C., & Kim, C. J. (2022). Using an ecological approach to explore teacher agency during the implementation of a citizen science education program using Arduino. *Asia-Pacific Science Education, 8*(2). https://doi.org/10.1163/23641177-bja10054

National Academies of Sciences Engineering and Medicine. (2018). *Learning through citizen science: Enhancing opportunities by design*. National Academies Press. https://doi.org/10.17226/25183

National Research Council [NRC]. (2013). *Next generation science standards: For states, by states*. The National Academies Press. https://doi.org/10.17226/18290

OECD. (2018). *The future of education and skills: Education 2030* (OECD education working papers). https://www.oecd.org/education/2030/E2030%20Position%20Paper%20(05.04.2018).pdf

OECD. (2023). *PISA 2025 science framework (DRAFT)*. https://pisa-framework.oecd.org/science-2025/assets/docs/PISA_2025_Science_Framework.pdf

Pelacho, M., Ruiz, G., Sanz, F., Tarancón, A., & Clemente-Gallardo, J. (2021). Analysis of the evolution and collaboration networks of citizen science scientific publications. *Scientometrics, 126*(1), 225–257. https://doi.org/10.1007/s11192-020-03724-x

Peter, M., Diekötter, T., Höffler, T., & Kremer, K. (2021). Biodiversity citizen science: Outcomes for the participating citizens. *People and Nature, 3*(2), 294–311. https://doi.org/10.1002/pan3.10193

Phillips, T. B., Porticella, N., Constas, M., & Bonney, R. (2018). A framework for articulating and measuring individual learning outcomes from participation in citizen science. *Citizen Science: Theory and Practice, 3*(2), 3. https://doi.org/10.5334/cstp.126

Phillips, T. B., Ballard, H. L., Lewenstein, B. V., & Bonney, R. (2019). Engagement in science through citizen science: Moving beyond data collection. *Science Education, 103*(3), 665–690. https://doi.org/10.1002/sce.21501

Preston, C., Golumbic, Y., Kenneally, C., Phway, T. S., Sousa, L. B., Martin, J., Rutledge, P., & Motion, A. (2023). Citizen science in schools – Perceptions of project leaders and teachers. *Teaching Science, 69*(3), 3–9.

Roche, J., Bell, L., Galvão, C., Golumbic, Y. N., Kloetzer, L., Knoben, N., Laakso, M., Lorke, J., Mannion, G., Massetti, L., Mauchline, A., Pata, K., Ruck, A., Taraba, P., & Winter, S. (2020). Citizen science, education, and learning: Challenges and opportunities. *Frontiers in Sociology, 5*, 110. https://doi.org/10.3389/FSOC.2020.613814

Salmon, R. A., Rammell, S., Emeny, M. T., & Hartley, S. (2021). Citizens, scientists, and enablers: A tripartite model for citizen science projects. *Diversity, 13*(7). https://doi.org/10.3390/d13070309

Sarmiento-Márquez, E. M., Pishtari, G., Prieto, L. P., & Poom-Valickis, K. (2023). The evaluation of school-university partnerships that improve teaching and learning practices: A systematic review. *Educational Research Review, 39*(January), 100509. https://doi.org/10.1016/j.edurev.2023.100509

Saunders, M. E., Roger, E., Geary, W. L., Meredith, F., Welbourne, D. J., Bako, A., Canavan, E., Herro, F., Herron, C., Hung, O., Kunstler, M., Lin, J., Ludlow, N., Paton, M., Salt, S., Simpson, T., Wang, A., Zimmerman, N., Drews, K. B., … Moles, A. T. (2018). Citizen science in schools: Engaging students in research on urban habitat for pollinators. *Austral Ecology, 43*(6), 635–642. https://doi.org/10.1111/AEC.12608

Schuttler, S. G., Sears, R. S., Orendain, I., Khot, R., Rubenstein, D., Rubenstein, N., Dunn, R. R., Baird, E., Kandros, K., O'Brien, T., & Kays, R. (2019). Citizen science in schools: Students collect valuable mammal data for science, conservation, and community engagement. *Bioscience, 69*(1), 69–79. https://doi.org/10.1093/biosci/biy141

Sforzi, A., Tweddle, J., Vogel, J., Lois, G., Wägele, W., Lakeman-Fraser, P., Makuch, Z., & Vohland, K. (2018). Citizen science and the role of natural history museums. In S. Hecker, M. Haklay, A. Bowser, Z. Makuch, J. Vogel, & A. Bonn (Eds.), *Citizen science: Innovation in open science, society and policy*. UCL Press. https://doi.org/10.2307/j.ctv550cf2.36

Shah, H. R., & Martinez, L. R. (2016). Current approaches in implementing citizen science in the classroom. *Journal of Microbiology & Biology Education, 17*(1), 17–22. https://doi.org/10.1128/jmbe.v17i1.1032

Silva, C. G., Monteiro, A., Manahl, C., Lostal, E., Holocher-Ertl, T., Andrade, N., Brasileiro, F., Mota, P. G., Sanz, F. S., Carrodeguas, J. A., Brito, M., & R. M. (2016). Cell spotting: Educational and motivational outcomes of cell biology citizen science project in the classroom. *Journal of Science Communication, 15*(1), 1–20.

Smith, H. E., Cooper, C. B., Busch, K. C., Harper, S., Muslim, A., McKenna, K., & Cavalier, D. (2023). Facilitator organizations enhance learning and action through citizen science: A case study of Girl Scouts' Think Like a Citizen Scientist journey on SciStarter. *Environmental Education Research*, 1–22. https://doi.org/10.1080/13504622.2023.2237705

Sparkling Science. (2020). Sparkling Science programme facts & figures. https://www.sparklingscience.at/_Resources/Persistent/69698fa1dcca33a8e3003b4bd15050f9527580fa/SPA_F%26F_2020_short_EN.pdf

Spicer, H., Nadolny, D., & Fraser, E. (2020). Going squirrelly: Evaluating educational outcomes of a curriculum-aligned citizen science investigation of non-native squirrels. *Citizen Science: Theory and Practice, 5*(1), 14. https://doi.org/10.5334/cstp.275

Tal, T., Barkae, H. S., & Alon, N. L. (2023). From school-based citizen science to transition driven activism with the community. In J. Diamond & S. Rosenfeld (Eds.), *Amplifying informal science learning* (pp. 141–154). Routledge. https://doi.org/10.4324/9781003145387-16

Trautmann, N., Manker, J., Mitchell, K., & Kahler, P. (2019). It's more than fluff: Using birds to build student interest and skills in science using birds to build student interest and skills in science on JSTOR. *The Science Teacher, 86*(5), 40–47. https://www.jstor.org/stable/26899114?casa_token=xol40rVYjUgAAAAA%3A5pFRurfQv4Y0SkazJJMHulcnvTzGs8Q4wGwV4uCM79XJiHRzjifTBqGNmLDdSa4gz2IzGqFLsFIOsBDgclel9jk6dQPzX4MW9Nb7ZyGIgvB0L3gG5lPE5Q&seq=1

Tsapalov, A., Kovler, K., Shpak, M., Shafir, E., Golumbic, Y., Peri, A., Ben-Zvi, D., Baram-Tsabari, A., Maslov, T., & Schrire, O. (2020). Involving schoolchildren in radon surveys by means of the "RadonTest" online system. *Journal of Environmental Radioactivity, 217*, 106215. https://doi.org/10.1016/j.jenvrad.2020.106215

Williams, K. A., Hall, T. E., & O'Connell, K. (2021). Classroom-based citizen science: Impacts on students' science identity, nature connectedness, and curricular knowledge. *Environmental Education Research, 27*(7), 1037–1053. https://doi.org/10.1080/13504622.2021.1927990

Zoellick, B., Nelson, S. J., & Schauffler, M. (2012). Participatory science and education: Bringing both views into focus. *Frontiers in Ecology and the Environment, 10*(6), 310–313. https://doi.org/10.1890/110277

Chapter 3
Promoting Expansive Framings by Design: How Citizen Science Can Foster Quality Education in Schools

Yael Bar Nes, Yael Kali, Ornit Sagy, and Amram Eshel

3.1 Introduction

In recent years our society has been rapidly changing. More than ever, we are facing environmental, health, economic and social challenges. These are major concerns for all, but especially for today's youths who will particularly need to cope with the consequences of these changes. Educators and educational researchers have been calling in the past decade to rethink what and how students are learning in schools, to equip them with knowledge, skills and mindsets that will assist them when facing such challenges. This is evident, for instance in the book "*What's Worth Teaching? Rethinking Curriculum in the Age of Technology*" (Collins, 2017), or more generally, in the United Nations' mission to ensure quality education by 2030 (SDGs, 2024), with a focus on lifelong learning opportunities for all. A global effort to distill key aspects of the much needed educational change, highlights the importance of nurturing students to be "active, responsible, and engaged citizens" (OECD, 2018, p. 4).

Such engagement is particularly important, given challenges faced by, and to a large extent caused by humanity in the Anthropocene. It is imperative that students will acknowledge human influence on our planet and develop at least a basic understanding of the complex mechanisms of this influence in order for them to act responsibly and intelligently towards sustainability. Such behavior has been termed Environmental Citizenship (Dobson, 2003), and its fostering through pedagogical means—has been termed Environmental Citizenship Education (Hadjichambis

Y. Bar Nes · Y. Kali (✉) · O. Sagy
Faculty of Education, University of Haifa, Haifa, Israel
e-mail: yael.bennun@edtech.haifa.ac.il; yael.kali@edtech.haifa.ac.il

A. Eshel
Faculty of Life Sciences, Tel Aviv University, Tel Aviv-Yafo, Israel

D. Tsybulsky, T. Tal (eds.), *Citizen Science Contributions to Biology and Environmental Education*, Contributions from Biology Education Research,
https://doi.org/10.1007/978-3-031-99298-8_3

et al., 2020). As elucidated in the introduction to this book, a promising direction for promoting environmental citizenship education is the incorporation of citizen science projects as part of science education in schools.

Citizen science is a genre of research where volunteer non-scientists collaborate with scientists to advance scientific knowledge (Hecker et al., 2018). This social phenomenon, which has roots that go back to Darwins' joint ventures with people who assisted him in tracking down and identifying local fossils (Cooper, 2016), was conceived and developed in arenas that originally did not include schools at all. However, informal education has gradually become a major aim of citizen science (Roche et al., 2020), and its potential for learning (Phillips et al., 2018) and schooling (Zoellick et al., 2012), are being progressively uncovered. In fact, school participation in citizen science—coined as SPICES—has recently been defined as a field of research in the learning and instructional sciences (Kali et al., 2023).

A particular focus of research in SPICES, given that many of the citizen science projects involve participants in environmental research, is the ways in which such involvement may promote the development of *environmental science agency* among students (Ballard et al., 2016). Environmental science agency "combines not only understanding of environmental science and inquiry practices, but also the youths' identification with those practices and their developing belief that the ecosystem is something on which they act" (p. 65). Research across various youth-centered community and citizen science projects, has shown that when youth understood how data they contributed, or knowledge they produced in citizen science projects was used by real scientists—they tended to develop a sense of responsibility and agentive learning (Harris et al., 2020). Such understanding of the youth represents an *expansive framing* of their activities in the project (Engle et al., 2012). As opposed to bounded framing, where learning is compartmentalized within specific contexts, the term expansive framing denotes contexts in which learners view themselves as "actively contributing to larger conversations that extend across time, places, and people" (Engle et al., 2012, p. 215). Expansive framing in citizen science tends to develop among learners typically in non-formal settings. In SPICES contexts however, students often stick with the bounded school perspective, where the purpose of learning is often perceived as succeeding in school and getting a good grade (Harris et al., 2020; Lemke, 1990). For example, such a perspective is evident in a student's answer regarding the importance of contributing data to a school-based citizen science project: "I need it [uploading a document with her data] to show that my name's on there so we would get a good grade." (Harris et al., 2020 p. 251).

Given its potential to foster environmental science agency, it is important to understand how design of citizen science curricular activities may support students' uptake of expansive framing in school contexts. The implementers of SPICES are typically entrepreneurial teachers and principals who partner with citizen science advocates outside the school, and are willing to adapt the citizen science activities to their own school contexts (Benichou et al., 2022, Kali et al., 2023). To enable such adaptations, SPICES curricula are often designed in a flexible-customizable manner (Sagy et al., 2020). The current study set out to explore how such flexible design may offer opportunities to foster expansive framings among students in various age levels in the context of biology and environmental education. We did so by analyzing the ways students framed their learning within different implementations of *The Bermuda Buttercup Mystery* initiative—a SPICES program designed to support curricular adaptations to incorporate the issue of invasive plant species into various educational contexts.

Our research is focused on two questions: (1) What characterizes students' framings of their learning in various implementations of *The Bermuda Buttercup Mystery Initiative*? (2) What is the relationship between students' framings, and the design features their teachers chose to implement?

3.2 Theoretical Background

This section provides a brief review of the literature on the idea of flexible and customizable design, and how it can equip teachers with knowledge and tools to adapt curricular materials to their own teaching contexts. Then, customizable design approaches for SPICES are reviewed to set the stage for the current study. Finally, current research regarding the potential of features within the design of SPICES programs to promote students' expansive framing of their learning is presented as a basis to answer the current research questions.

3.2.1 Customizable Design

When designing curriculum materials, especially when the design takes place within research-practice partnerships, and in design-based research (DBR) approaches (see Methods), one must be careful not to fall into the trap of designing "boutique" projects, which are applicable to the specific contexts for which they were designed, but with little adaptability to additional contexts (Kidron & Kali, 2017). Dimitriadis and Goodyear (2013) propose strategies to broaden the scope of DBR projects by: (a) acknowledging the intricacies of implementing the design in diverse settings; (b) anticipating adjustments teachers may make to tailor the design to new environments; (c) assisting teachers in implementing the innovation beyond the research setting; and (d) periodically assessing and refining the innovation. Consequently, they claim that good design should anticipate the need for modifications and adaptations to meet various types of specific requirements ("design for configuration"), and offer real-time support for teachers' implementation efforts ("design for orchestration"). Designing for flexibility employs such strategies, capitalizing on a conceptualization that views teaching as a design profession (Kali et al., 2015; Laurillard, 2013). In doing so, it acknowledges teachers' important role in making adaptations, which ultimately open up educational innovation to a broad range of learners in various contexts.

3.2.2 Design for SPICES

Flexible and customizable curricular design is especially required when SPICES contexts are concerned (Kloetzer et al., 2021; Thiel et al., 2023). A main reason is that citizen science projects are typically designed outside the school arena with agendas that are quite different and sometimes compete with school agendas (Sagy et al., 2019; Atias et al., 2023). Although citizen science projects often intertwine educational goals with the aim of advancing science, the latter does not necessarily comply with schooling practices and norms. Therefore, guidelines for designing citizen science projects, such as the ten principles for quality citizen science (Robinson et al., 2018), and even guidelines aimed at fostering learning within out-of-school citizen science endeavors (Herodotou et al., 2018; Kloetzer et al., 2021; Pandya & Dibner, 2018) can only serve as a basic layer in SPICES. Another layer of design guidelines that will support the school-academia interface is still a much needed research area to advance knowledge and practice in the young field of SPICES.

Despite the considerable potential of citizen science to boost science and environmental education with interesting and engaging authentic contexts, it is important to acknowledge a notable lack of alignment between the contents that citizen science projects offer and typical school syllabi (Kloetzer et al., 2021; Matuk et al., 2023). Citizen science projects include contents that are typically far beyond, but at the same time do not cover such syllabi. Moreover, at times, a single citizen science project can address a variety of disciplinary topics, affording its incorporation into various school subject matters or serve as basis for interdisciplinary syllabi. This makes it crucial for the design of activities in SPICES learning materials to be flexible enough to facilitate the adoption and implementation of the citizen science contents across a range of educational objectives and for diverse learning contexts. At the same time, it is imperative to provide learning-design tools that will assist teachers in adapting these activities to local constraints and affordances, as well as support their careful balancing of the embedded educational and scientific goals (Law et al., 2024).

3.2.3 Design for Expansive Framing in Contributory SPICES

Another aspect of flexibility in the design of SPICES has to do with the characteristic of participation anticipated in the citizen science project incorporated. Within citizen science projects, various modes of participation have been identified, each offering distinct approaches of engagement (Bonney et al., 2009): contributory projects involve volunteers in collecting data following predefined protocols; collaborative projects engage participants in various stages of the research process, fostering a sense of partnership between scientists and volunteers; and co-created projects emphasize collaboration from the project's inception, with volunteers and researchers jointly designing research questions and methodologies. Each of these engagement modes offers different types of opportunities and challenges for designing their incorporation into school contexts.

Contributory projects, which are in general the more common mode of citizen science, are also those that are often chosen for SPICES initiatives (Hecker et al., 2018; Kloetzer et al., 2021), as in the case of the current study. Recent research has shown that despite being considered as inferior with respect to their potential to support in-depth student learning, contributory projects may encourage a meaningful path for SPICES, embodying students' and teachers' responsibility over scientific processes and outcomes (Atias et al., 2023). However, not much is known about whether contributory citizen science projects that are embedded in SPICES may foster expansive framings among students, especially in the context of environmental education, and what types of design features may support such framings.

3.3 Methods

3.3.1 Methodological Approach

The study was conducted as a first, exploratory iteration of a DBR. DBR is a methodological approach in which theoretical conjectures regarding learning are systematically tested by designing curricular activities and iteratively implementing them with learners in the context for which the conjectures are made. Each iteration enables the researcher to draw tentative conclusions about the relationship between the design and learning processes and outcomes, and make refinements in the designed features to further test these assumptions in the additional iterations. Preliminary iterations are usually exploratory in nature, as in the current research, and additional iterations tend to provide further generalizability. In this way DBR serves two complementary goals of advancing both theoretical knowledge about various aspects of learning within specific contexts, as well as design knowledge about how to support such learning (Barab & Squire, 2004; DBRC, 2003; Kali and Hoadley 2021; McKenney & Reeves, 2018), as sought in this research. To conduct this exploratory stage we employed a multiple case-study approach (Stake, 2013).

3.3.2 The Citizen Science Project at the Heart of **The Bermuda Buttercup Mystery** *Initiative*

The Bermuda Buttercup Mystery initiative is a collaboration between school students and a plant scientist (the fourth author of this chapter). The citizen science project at the heart of this initiative seeks to investigate the rapid spreading of *Oxalis pes-caprae*—an invasive species covering extensive areas in Israel, and disrupting the sprouting and growth of native species. Based on a similar phenomenon found in Portugal, the scientist hypothesized that a new type of *Oxalis pes-caprae* (type M or type L in Fig. 3.1), with a different floral morph than the abundant type (type S), had invaded Israel, and enabled sexual reproduction, causing accelerated spreading (Eshel, 2023). As shown in Fig. 3.1, in type S—the Short styled morph—the stigma are at the bottom, below two layers of stamens, in type M—the Mid styled morph—the stigma ar at the middle, between the two layers of stamens and in type L—the Long styled morph—the stigma are at the top, above the two layers of stamens. To examine this hypothesis, the scientist developed a citizen science protocol enabling non-experts to easily participate in the project by contributing data. Participation in the project requires participants to (a) remove the petals of the flower to reveal the stamens and stigmata, (b) identify the type of the flower (S, M or L) and (c) report the finding into a shared online map, including the location of the sampling, and a picture showing the reproductive parts of the flower.

To encourage participation in the project, the scientist approached the *Taking Citizen Science to School* (TCSS) center, whose mission is to develop and nurture a network of partnerships between schools and scientists through ongoing citizen science projects (TCSS, 2024). The TCSS modus operandi is based on co-designing and adaptation of instructional materials in multi-expertise teams within the network, which connect the contents of citizen science projects in various topics with school national syllabi. One criterion for choosing these projects is that the partnership with schools will continue for several years. While implementing these projects in classrooms, TCSS provides ongoing support for teachers and schools, and invites them to take part in the co-design and adaptation process. The network of partnerships also functions as a learning organization, exploring the notion of SPICES (Aridor et al., 2022; Kali et al., 2023; Sagy et al., 2020). The *Bermuda Buttercup Mystery initiative* was conceived when the first author of this chapter took up the

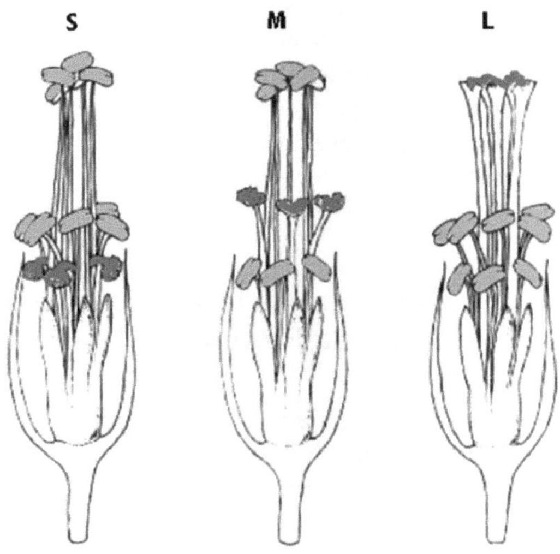

Fig. 3.1 Illustration of the three floral morphs of *Oxalis pes-caprae*: S-Short styled, M-Mid styled, L-Long styled. (Adapted from Castro et al., 2007)

Oxalis pes-caprae project as her graduate level research topic within TCSS, with the supervision of the second and third authors. As part of her research, she led the design of the curriculum materials, while supporting teacher adaptation and implementation.

3.3.3 Sequence of Activities and Main Design-Features

To examine students' framings of their learning in *The Bermuda Buttercup Mystery* initiative, and explore how such framings may relate to the design of the implementation they took part in, we developed a sequence of curricular activities for modular and flexible use by teachers with their students. To do so we used *EXPANSION*—a digital platform developed in the TCSS research center (TCSS, 2024; Law et al., 2024), in which this study took place. The platform provides teachers with readily available curriculum materials for incorporating citizen science projects (such as the *Oxalis pes-caprae* project in the current study) into their own teaching. For each project resources are offered, which teachers can use for guiding their students as they take part in scientific research, optionally exploring their own research questions. The set of resources is organized as a sequence of activities that teachers can customize using planning and design tools that assist teachers while exploring ways to connect the project with local characteristics, constraints, and affordances, in a teachers-as-designers approach (Kali et al., 2015). Sequences are typically organized around (a) introduction activities that present the problem and set the stage for (b) a set of activities in which students take part in the citizen science project, with optional extended inquiry activities that guide students as they conduct a small empirical research on the topic, and (c) reflection activities, where students ponder on their learning, their participation in the scientific research, and their responsibility as citizen scientists. Figure 3.2 illustrates the sequence of activities offered for teachers' customization in *The Bermuda Buttercup initiative* (Each circle represents one lesson. Clicking the circle offers curricular materials and resources for teachers including the design principles that guided the original design of each activity and should be considered in adaptations).

Within *The Bermuda Buttercup Mystery* sequence, we encapsulated four main types of common design-features, each focusing on an important element of participation in citizen science projects. These consist of: The **nature of participation** in authentic research in general, and in the specific citizen science project in particular (DF1); **Acquaintance with the scientists** in the project (DF2); **Gamification of data sampling and reporting**, e.g., using competitions, scoring rubrics, etc. (DF3); and **Extended inquiry** beyond the basic protocol (DF4). Table 3.1 illustrates how these design-features were embodied in the sequence of activities.

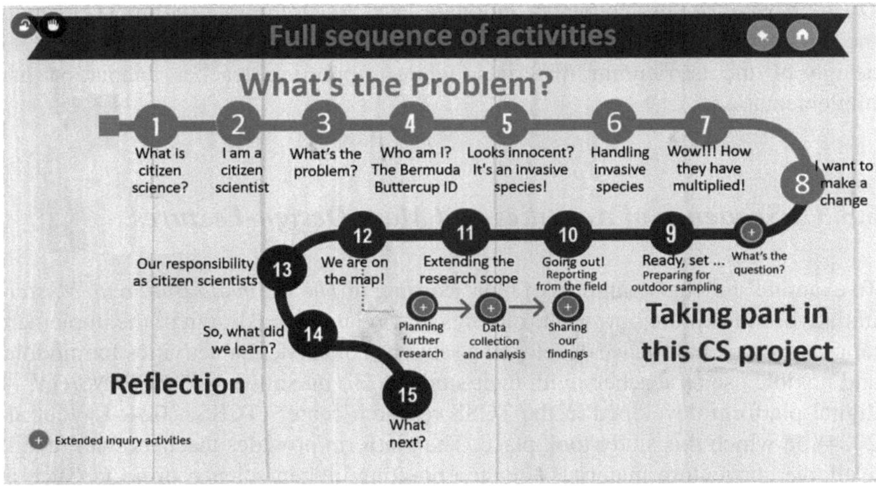

Fig. 3.2 Sequence of activities offered for teachers' customization in the Bermuda Buttercup Mystery initiative in the *EXPANSION* digital platform

Table 3.1 Types of design-features embedded with sequence of activities

Section within sequence	Types of design-features embodied in section
Introduction to problem, scientific concepts involved, and the citizen science protocol for participation (activities 1–8)	DF1, DF2
Participation in the citizen science project (activities 9–12)	DF1, DF2, DF3
Extended inquiry (additional activities marked +)	DF4
Reflection (activities 13–15)	DF1

3.3.4 Participants

Eighty two students who participated in The Bermuda Buttercup Mystery initiative in three different classes (grades 5, 8 and 12) from three schools located in various geographical areas in Israel participated in the study along with their teachers. Each class served as a case study. The cases were chosen out of 11 implementations (about 340 students who studied in 14 classes in 11 schools) of the initiative during the year 2023. The rationale for choosing the three cases was to represent different types of design-features that the teachers emphasized in them, as well as different age levels of the students, as illustrated in Table 3.2.

Table 3.2 Types of design-features employed in each case study

Case	Grade and no. of students	DF1 Nature of participation in CS project	DF2 Acquaintance with the scientists	DF3 Gamification of data sampling and reporting	DF4 Extended inquiry
A	Grade 8 26 students	V	V	V	
B	Grade 5 30 students	V	V		
C	Grade 12 26 students	V			V

3.3.5 *Data Collection and Analysis*

In each of the schools an in-depth semi-structured interview was conducted in a focus group of 4–5 students, who were chosen by their teacher as representing a wide range of abilities and interest in the project. The interviews, which lasted about 40 min each, and conducted at the end of each implementation, were based on interview questions in studies regarding environmental science agency, where expansive framing plays an important role (Ballard et al., 2016), which were slightly adapted for the current research. Students were asked to describe their participation in the initiative, to reflect on what they learned, the value they saw in their participation, and actions they took or want to take in the future with regard to the project. All interviews were audio-recorded and transcribed. In addition to the interviews, informal discussions with the teachers enabled us to characterize the various implementations and the design-features emphasized in each implementation (as outlined in Table 3.2). Details regarding the sampling (number and place) were obtained from the project's website, and corroborated with information elicited from the discussions with the teachers. Finally, student artifacts created during various parts of each implementation were also gathered (e.g., brochures created by students to invite others to contribute to the scientific research in Case A, or posters created as part of the extended inquiry to share findings, in Case C). Data were analyzed qualitatively, using inductive analysis, with an eye on the notion of expansive framings.

3.4 Findings: How Expansive Framings Emerged in Three Different Implementations

3.4.1 *Case A*

This case took place in a class with 26 eighth-grade students with a specific interest in science in a school for girls. In addition to implementing the full set of activities, the teacher decided together with her students to put a specific effort in contributing a large set of samples of high quality to support the project. To do so, she: (a) invited the project leader (first author of this chapter) to meet with the students for the opening session of the initiative, introduce the problem, explain how the students can take part in the research, and demonstrate how to carry out the sampling and reporting protocol (DF1, DF2); (b) carried out several field work sessions with the students, to assure that they understood and were proficient in conducting the citizen science protocol of removing the petals to reveal the stamens and stigmata, and identifying the type of *Oxalis pes-caprae* (strengthening DF1); (c) added a gamified aspect to the data collection process by developing a scoring rubric where teams of students gained extra points to their grade as they reported on greater numbers, wider geographical areas, and higher quality (e.g., good pictures and descriptions) of samples uploaded to the shared map on the project's website over the course of the project (DF3); (d) implemented an additional activity where students developed brochures to invite others to contribute *Oxalis pes-caprae* samples to the project's website, send them to their family and friends who live in other areas of the country, and support them in this process (strengthening DF3).

Figure 3.3 illustrates one example of a brochure produced by a group of students in Case A, 'marketing' the project in response to the competition their teacher initiated (DF3). As a result of the emphasis on quantity and quality of data collection to support the project's research in this implementation, students in this case were

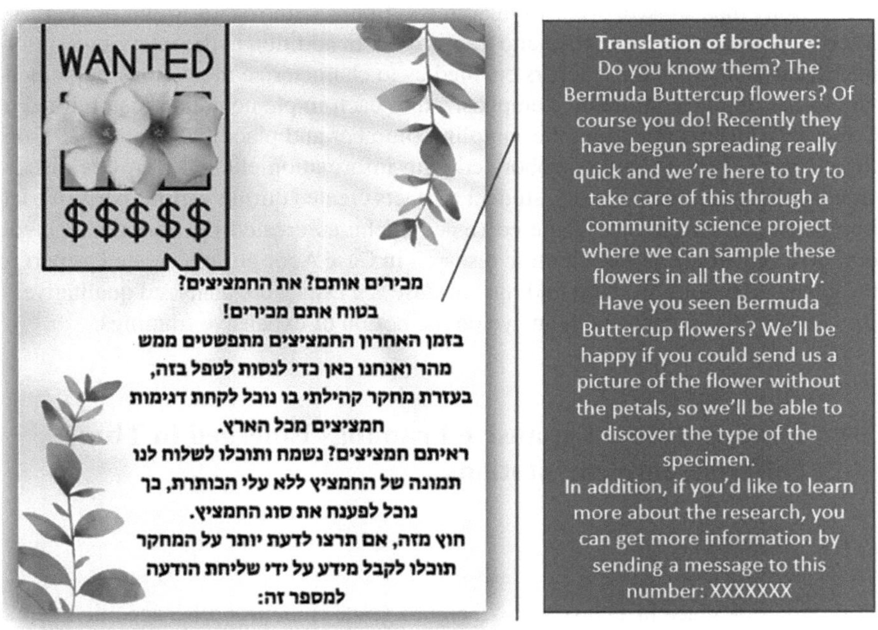

Fig. 3.3 Example of a brochure (in Hebrew, with a translation added) developed by one group of students in Case A

responsible for uploading the largest amount of quality samples—84 in total, from 13 different locations in Israel—to the shared online map. This was much more than the sampling contributed by any of the classes who participated in the initiative.

In the analysis of the focus group interview with students from this class we found evidence for students' development of environmental science agency, which seems to be related to an expansive framing that resonates with the design-features emphasized in this implementation. For example, one student explained how inviting other people to participate in the project, which was fostered by the teacher's implementation of DF3—the design feature of gamification of data sampling and reporting—nurtured in her a sense of responsibility and agency in the project:

> *There are now* [following the advertising efforts students made, via their teacher's implementation of DF3] *many students from all over the country in this research, and we're all here now, searching for Bermuda Buttercup specimens, marking them on the map, and brainstorming ways to stop their spread. And that's important because if we want other plants to remain in the world, we need to stop the spread of this invasive species…* ***I felt like I was part of something much bigger that is really important.***

We view this type of expansive framing as an "ambassadorship framing". It reflects students' wish to recruit others to participate in the project due to its importance to science or the environment.

Another student emphasized the value of choice, which was part of the "rules of the game" the teacher came up with in her scoring rubric and voluntary participation in the competition. Apparently, this played an important role in students' agency to take part in the project and invite others to do so:

> *I think the most meaningful lesson was the one where the teacher said we were going to split into groups and start looking for the Bermuda Buttercup specimen, because then I really understood that…* ***the option was in my hands, if I want to take part in this or not****, and at that moment I realized that this topic mattered to me. I want to succeed in it. I want to help other people,* ***I want to tell other people what I know…*** *It was a lesson where I understood that I really want to help the environment and be part of this initiative of the Bermuda Buttercup citizen science.*

The ambassadorship type of framing is evident in the above excerpt, in the student's wish to tell other people about the project and its importance to the environment.

Environmental science agency also seemed to be related to an expansive framing with a personal flavor, of helping not only the environment, but also people who care about it, for instance, in the following example where a student conveyed her wish to "make a change" by helping the initiative leader and "other experts":

> *In addition to taking part in the Bermuda Buttercup research that made us see how impactful it was,* ***it made me want to make a change*** [in the environment]. *We really did amazing things. We found Bermuda Buttercup flowers in so many places…* ***I knew I was helping you*** **[the initiative leader]** ***and other experts****, and it really gave a good feeling.*

We view this case as representing an expansive framing we call "helping the scientist framing".

To summarize, in this case the teacher implemented design features DF1, DF2 and DF3. Environmental science agency was related to expansive framings of two categories: an ambassadorship framing and a "helping the scientist" framing.

3.4.2 Case B

The collaboration with this school was different from all the other schools. The teacher did not participate in any of the TCSS meetings. Rather, he learned about the *Oxalis pes-caprae* project in a professional development program for Biology teachers, where the scientist presented the project. He decided to implement the program on his own, without any support from *The Bermuda Buttercup Mystery* initiative leader, and was not aware of the curriculum materials offered. Since outdoor learning was greatly valued in the elementary school where he taught, he had much experience teaching outdoors and felt comfortable to introduce the project's research questions and protocol to his fifth grade students, and take them out to the field, accompanied by several other teachers to conduct some sampling with them. He also explained the notion of citizen science, so that students were aware of the partnership with the scientist and the purpose of their sampling, thereby implementing the *Nature of participation in citizen science design-feature* (DF1).

To everyone's surprise, one of the students found a specimen that he identified as type M—*Oxalis pes-caprae*. That same day the teacher found the initiative leader's contact information on the TCSS website, and informed her about the exciting discovery. The scientist drove to the school to meet the student and the teacher, and examined the specimen (DF2). Indeed, he confirmed that the first M type *Oxalis pes-caprae* was found by the student, who helped confirm the hypothesis that this type exists in Israel, and might be responsible for the rapid proliferation of the species in the country. The finding, and the story of how it was found, including a picture of the scientist shaking hands with the student was published in a national website devoted to the study of wild plants in Israel (Eshel, 2023), and also in a local newspaper.

Here too, the characteristics of expansive framings that were evident from statements in the focus group interview, resonated the learning experiences students went through.

First, many students expressed their pride not only in being a contributing part of the grownup world, but also in having an added value in such contribution:

> *This proves that kids, if they try hard enough... can help in things that people think that only grownups can do. It also proves that* ***kids can do many things that grownups don't always succeed in doing...*** *We have small hands that don't cover our view and we have a sharper vision.*
>
> *Second, they tended to frame their contribution to the project as a partnership between kids and scientists, where kids have an important role in advancing science:*
>
> *We find the problem (the "problematic" type M specimen) and they (the scientists) examine and further research the problem. We find and the scientists research...* ***it's a great combination that will help [advance] science.***

We view this type of expansive framing as an "advancing knowledge framing", reflecting students' sense of capacity to take part in the doing of science.

Additional statements illustrate the ambassadorship expansive framing found also in Case A. In this case, this was due to their group's finding of the M-type specimen:

> *This is something that we need to tell other people about.* ***We made an important scientific finding and so can others.*** *This way they will also be able to help [advance science] and tell more people, and this will go on and on.*

To summarize, in this case the teacher implemented only design features DF1 and DF2. Environmental science agency was related to expansive framings of two categories: an "advancing knowledge framing" and an ambassadorship framing.

3.4.3 Case C

The implementation of *The Bermuda Buttercup Mystery* in this case took place with 26 12-grade biology-major students who took the program as their "Bio-inquiry" project. Bio-inquiry is a guided inquiry project that constitutes 20% of the final grade for biology majors in Israel. As described in the Ministry of Education website, students in these projects are required to develop a deep understanding of a specific biological phenomenon of their interest through planning and conducting an experiment regarding a research question they come up with following a literature review.

With the aid of the initiative leader, the teacher in this case adapted the suggested sequence of activities to meet the scope and requirements of a Bio-inquiry. Specifically, she emphasized and expanded the *Extended inquiry* (DF4) activities in the sequence (blue circles in Fig. 3.2) to support her students in conducting their own research. To do so, she interacted with the scientist to understand what type of questions can potentially advance the knowledge regarding *Oxalis pes-caprae*, beyond the reproduction question in the project's protocol, and beyond what is currently known. Since research on this species is scarce, the scientist suggested several directions, which eventually students used as a basis for their inquiry projects. For example, one group explored the effect of various types of covering, and another studied the effect of pine needle oil and the growth of the *Oxalis pes-caprae* plants. All these were conducted in an area within the school's yard where these plants were present (Fig. 3.4). It is important to note, though, that except for asking his advice on directions for research questions, the teacher did not ask for further assistance from the scientist, nor did she share her students' findings with him.

In addition to the Bio-inquiry projects, the teacher implemented the whole sequence of activities (encapsulating DF1 and DF2, as shown in Figure). However, eventually, since the extended inquiry took much time and focus, students did not report on the sampling and identification they conducted with specimens they

Fig. 3.4 Students in Case C working on their Bio-Inquiry projects (left and middle), and specimen brought to the classroom as part of the inquiry

brought to the classroom from both the school's backyard, and three other places in Israel.

Interestingly, despite the fact that these students did not contribute to the citizen science project—the analysis of this case indicates that they seemed to have framed their inquiry expansively. That is, even though their findings were not shared with anyone beyond the classroom, all students in the focus group expressed excitement about conducting research in areas unknown to scientists, with what they perceived as having important implications to the relations between society and the environment. Their environmental science agency seemed to be related to an "advancing knowledge framing":

> *We found that under the pine trees the Bermuda Buttercup doesn't grow, and in the other group they found other ways to slow down or even stop their growth. It was worthwhile examining this.* ***Our knowledge can help the farmers*** *handle things differently. They can find a way to slow down or even stop their [Oxalis pes-caprae] growth in the crop fields. In the end we all benefit from this.*

"Helping the scientist framing" was also evident:

> *In the long run,* ***we contribute [knowledge] to the professor,*** *and the professor contributes to the farmers.*
>
> Or
>
> *In this project you contribute to real research and* ***real researchers who need this help.***

However, some of the students understood that their contribution to advancing science, despite being meaningful, is also limited, because they did not have a deep conversation with the scientist. This was conveyed for instance, in the true desire for such a conversation by this student:

I would have liked to share the findings with Amarm [the scientist], because he seems to be the most relevant person to discuss it with. **I will give him my findings and he will give me the reasons. He will explain to me the logic behind them**: things I don't know, and things that are more accurate. And I think it's a conversation I would be happy to have. Really, to get a better and more professional perspective on what I'm doing. Or for example, **to receive critique on our experiment, and how we could improve it**... It's reasonable to assume that if I talk to him about certain topics, he will bring a different perspective... I think that's really the beauty of citizen science... The more people, the more angles...

3.5 Discussion

Given the potential of expansive framing to foster environmental science agency among learners (Engle et al., 2012; Harris et al., 2020), this section discusses the emergence of expansive framings as presented in the three case-study findings. Furthermore, the relationship between such framings, and the design features teachers chose to implement in these cases is elucidated.

Across the cases, three main categories of expansive framing emerged from the analysis: an ambassadorship framing (Cases A and B), a "helping the scientist" framing (Cases A and C), and an "advancing knowledge" framing (Cases B and C). These categories echo similar characteristics of expansive framings found among youth in out-of-school contexts (Harris et al., 2020). We argue that the expansive framings found in the current study within formal school contexts are related to the different emphases embedded in each implementation (the various combinations of design features from DF1 to DF4), which the flexible design of *The Bermuda Buttercup Mystery* had enabled, as we explain below.

3.5.1 An Ambassadorship Framing

I want to tell other people what I know (Case A)

We made an important scientific finding and so can others (Case B)

One of the themes that emerged from the way students framed their learning in the project was the issue of ambassadorship. This theme was particularly prominent among students in Case A. It seems that students in this case were excited about being able to significantly contribute to the research at hand. The *Gamification of data sampling and reporting* design-feature (DF3) was important in leading students to happily take on the role of ambassadors for the project. The brochure-development strategy, together with the scoring rubric, and the element of choice that the teacher came up with, were a powerful way to implement this feature. It appears that taking on the role of ambassadors was exciting and joyful, and strengthened students' framing of the experience as contributing to a larger endeavor. The task of increasing participation in the project seemed to have developed a sense of responsibility and agency among the students in this case, as well as a recognition of the collective value of the multiple perspectives required to advance scientific understanding of the phenomenon.

Interestingly, the theme of ambassadorship also appeared, although somewhat less prominently, among students in Case B. It seems that in this class, although only the *Nature of participation in citizen science* (DF1) and *Meeting the scientist* (DF2) design-features were implemented—the very finding of the M-type *Oxalis pes-caprae*, and the fame it brought with it—created a sense of mission and a desire to spread the message to the world, and invite others to take part in the project.

That is, inherently, citizen science has the potential to develop in students a sense of ambassadorship, which in turn, may encourage deeper engagement with the content and the practices required for carrying out the protocol. This potential can be harnessed through the design-features of *Gamification of data sampling and reporting* (DF3), which not only lead to an increased contribution to the project (contribution of more data) but also to students' broader framing of their learning in a way that fosters agency to act in favor of the environment.

3.5.2 A "Helping the Scientist" Framing

> *I knew I was helping you and other experts and it really gave me a good feeling* (Case A)

> *In this project you contribute to real research and real researchers who need this help* (Case C).

Another theme that emerged regarding the way students viewed their learning in the project is a "helping the scientist framing." In this framing, students tended to view their learning not only for their own benefit, but also for the benefit of a knowledgeable person who can use the outcome of their learning in a way that would serve society and the environment. This theme was particularly prominent among students in Cases A and C. In Case A, the emphasis on extensive sampling and reporting, which created the sense of ambassadorship mentioned above (DF3), may have also led students to acknowledge the clear difference in the roles of the scientist and participants in contributory citizen science projects. This might have influenced their framing of the learning as "helping the scientist". It is worth noting that students' testimonies in both Case A and Case C indicate that they were excited to help, felt pride in partnering with the scientist, and enjoyed what they perceived as the derived prestige. Furthermore, helping the scientist was perceived as empowering, when students viewed their participation in the project as a response to the scientist's call for their assistance (as the quote above from Case C indicates).

The evidence for the "helping the scientist" framing in Case C is somewhat surprising. That is because students in Case C, where the *Extended inquiry* was emphasized (DF4) did not meet the scientist (no implementation of DF2) at all. Additionally, students in Case C did not upload their samples to the collaborative map, nor did they share their Bio-inquiry findings with the scientist (limited implementation of DF1). In other words, they did not actually help the scientist. This point is also discussed in the context of the next theme of an "advancing knowledge" framing, which was prominent in Case C.

It seems, therefore, that the *Gamification of data sampling and reporting* (DF3) and the *Extended inquiry* (DF4) design-features contributed, each in its own way, to the development of an expansive framing of learning, in which students perceived their learning in *The Bermuda Buttercup Mystery* as helping the scientist. Such framing was often perceived as a win-win partnership, where the scientist benefits from students' assistance in collecting data, and the students gain the opportunity to engage with exciting contents and practices required to partner with the scientist, as well as potential prestige.

3.5.3 An "Advancing Knowledge" Framing

> *We find and the scientists research… it's a great combination that will help* [advance] *science* (Case B)

> *Our knowledge can help the farmers handle things differently* (Case C)

The third theme that emerged from the analysis is perhaps the one that is most expected in SPICES programs—an "advancing knowledge" framing. In this framing students tended to view their learning as potentially advancing scientific knowledge regarding the rapid spread of the *Oxalis pes-caprae*, and practical knowledge regarding ways that people, such as farmers, may deal with this issue. Not surprisingly, this framing was especially prominent among students in Case B, where the type-M specimen was found.

Interestingly, this framing was also prominent among students in Case C, where the *Extended inquiry* design feature (DF4) was emphasized. Research has shown that when educators foregrounded research questions that students came up with in SPICES initiatives, such as in the Bio-Inquiry projects in Case C, students tended not to expansively frame their participation in the project (Harris et al., 2020). We suggest that students' development of an "advancing knowledge" framing in the current case was influenced by their teacher's consultation with the scientist regarding research areas with potential to advance scientific knowledge. Even though they were aware that their research findings did not eventually reach the scientist, it is possible that the expansive framing students developed stemmed from their understanding that they are researching an uncharted scientific territory.

Therefore, it is imperative that when implementing the *Extended inquiry* design-feature (DF4)—the state-of-the-art in the field is accessible to students, including what is known and what is yet to be discovered. For example, in the current study, the project leader updated all the teachers when the M-type *Oxalis pes-caprae* was discovered. New research questions were raised and updated in the *EXPANSION* digital platform, such as: How common and spread out is this morph? Can its invasion pathway be tracked? and more.

That said, it is worth dwelling on the words of the student who expressed his desire for more meaningful interaction with the scientist. In fact, this desire highlights an unrealized potential in the *Extended inquiry* design-feature of *The Bermuda Buttercup Initiative*, where a more collaborative mode of participation (see for example Tal et al., 2023) could have been added into this contributory mode of citizen science project. In the current case, not much would have been needed to realize this potential, since the scientist was interested in providing feedback to students on their research design and engaging in dialogue regarding its findings and implications. Such interaction may have enabled collaborative advancement of scientific knowledge regarding the *Oxalis pes-caprae*. Adding feedback sessions to the sequence of activities could have encouraged the teacher, who may have felt uncomfortable to do so otherwise, to continue the dialogue with the scientist and realize this potential. We suggest that although contributory, collaborative, and co-created modes of participation in citizen science projects are often clearly distinguished (Bonney et al., 2009), the current research indicates the potential for incorporating an optional collaborative trajectory of activities within a contributory citizen science project.

3.6 Conclusion

With the ultimate goal of promoting environmental citizenship among students, the present study aimed to examine how flexible design of SPICES projects can provide avenues for promoting expansive framings among students within the realms of biology and environmental education. Our findings indicate that indeed, SPICES projects, even when implementing a contributory mode of participation, inherently have the potential to encourage students' expansive framings across different implementations and age groups.

Making explicit *the nature of participation in the citizen science* project (DF1), which was implemented in the three case studies, and *acquaintance with the scientist* (DF2) are two basic design-features that play an important role in fostering expansive framings in SPICES contexts. Even students who participated in Case B where only these two features were implemented developed both an ambassadorship and an advancing knowledge expansive framings. The ambassadorship framing can be strengthened by *gamification of the data sampling and reporting* (DF3), as was found in Case A, where students did not even meet with the scientist. Regarding the advancing knowledge framing, the findings indicate that it can be strengthened by the *extended inquiry* design-feature (DF4), provided that the inquiry is conducted around research areas that have the potential to advance science, as was found in Case C. This case also highlighted the potential of the extended inquiry to provide an optional trajectory of adding a collaborative mode of participation into a contributory citizen science project.

Further research is required to continue developing the SPICES field's understanding of the relationship between types of design features that can be implemented in various ways in schools, and expansive framings that such features may promote. However, this research's findings strengthen the growing evidentiary basis for the call to adopt SPICES more systematically within educational systems as a strategy to advance environmental citizenship education in the spirit of the United Nations' mission of nurturing students as active, responsible and engaged citizens.

Acknowledgments This research was conducted as part of the TCSS Center of Excellence, which was supported by the Israel Science Foundation (grant no. 2678/17), and the Israel Science Foundation's MAPATS program (grant No. 991/23).

References

Aridor, K., Shaham-Dolev, R., Lavie-Alon, N., Valin, G., Sagy, O., & Kali, Y. (2022). Coalescing knowledge within networks of research practice partnerships: The case of a school-based citizen science network. In *Proceedings of the 16th International Conference of the Learning Sciences-ICLS 2022* (pp. 2114–2115). International Society of the Learning Sciences.

Atias, O., Kali, Y., Shavit, A., & Baram-Tsabari, A. (2023). Meaningful participation of schools in scientific research through contributory citizen science projects. *Science Education, 107*(5), 1163–1192.

Ballard, H. L., Dixon, C. G. H. H., & Harris, E. M. (2016). Youth-focused citizen science: Examining the role of environmental science learning and agency for conservation. *Biological Conservation, 208*, 65–75.

Barab, S. A., & Squire, K. (2004). Design-based research: Putting a stake in the ground. The *Journal of the Learning Sciences, 13*(1), 1–14. https://doi.org/10.1207/s15327809jls1301_1

Benichou, M., Kali, Y., & Hod, Y. (2022). Teachers' expansive framing in school-based citizen science partnerships. In *Teacher learning in changing contexts* (pp. 256–276). Routledge.

Bonney, R., Cooper, C. B., Dickinson, J., Kelling, S., Phillips, T., Rosenberg, K. V., & Shirk, J. (2009). Citizen science: A developing tool for expanding science knowledge and scientific literacy. *Bioscience, 59*(11), 977–984.

Cooper, C. (2016). *Citizen science: How ordinary people are changing the face of discovery*. Abrams.

Collins, A. (2017). *What's worth teaching? Rethinking curriculum in the age of technology*. Teachers College Press.

Castro, S., Loureiro, J., Santos, C., Ater, M., Ayensa, G., & Navarro, L. (2007). Distribution of flower morphs, ploidy level and sexual reproduction of the invasive weed Oxalis pes-caprae in the western area of the Mediterranean region. *Annals of Botany, 99*(3), 507–517.

Design-Based Research Collective (DBRC). (2003). Design-based research: An emerging paradigm for educational inquiry. *Educational Researcher, 32*(1), 5–8.

Dimitriadis, Y., & Goodyear, P. (2013). Forward-oriented design for learning: Illustrating the approach. *Research in Learning Technology, 21*(Suppl 1), 1–13. https://doi.org/10.3402/rlt.v21i0.20290

Dobson, A. (2003). *Citizenship and the environment*. OUP Oxford.

Engle, R. A., Lam, D. P., Meyer, X. S., & Nix, S. E. (2012). How does expansive framing promote transfer? Several proposed explanations and a research agenda for investigating them. *Educational Psychologist, 47*(3), 215–231.

Eshel, A. (2023, October 7). A sad announcement. *Wildflowers of Israel*. Retrieved March 1, 2024, from https://www.wildflowers.co.il/hebrew/ReadMore.asp?ID=844

Hadjichambis, A. C., Reis, P., Paraskeva-Hadjichambi, D., Činčera, J., Boeve-de Pauw, J., Gericke, N., & Knippels, M. C. (2020). *Conceptualizing environmental citizenship for 21st century education* (p. 261). Springer Nature.

Harris, E. M., Dixon, C. G., Bird, E. B., & Ballard, H. L. (2020). For science and self: Youth interactions with data in community and citizen science. *Journal of the Learning Sciences, 29*(2), 224–263.

Hecker, S., Haklay, M., Bowser, A., Makuch, Z., Vogel, J., & Bonn, A. (2018). *Citizen science: Innovation in open science, society and policy*. UCL Press.

Herodotou, C., Aristeidou, M., Sharples, M., & Scanlon, E. (2018). Designing citizen science tools for learning: Lessons learnt from the iterative development of nQuire. *Research and Practice in Technology Enhanced Learning, 13*, 1–23.

Kali, Y., & Hoadley, C. (2021). Design-based research methods in CSCL: Calibrating our epistemologies and ontologies. In *International handbook of computer-supported collaborative learning* (pp. 479–496). Springer International Publishing.

Kali, Y., McKenney, S., & Sagy, O. (2015). Teachers as designers of technology enhanced learning. *Instructional Science, 43*, 173–179.

Kali, Y., Sagy, O., Matuk, C., & Magnussen, R. (2023). School participation in citizen science (SPICES): Substantiating a field of research and practice. *Instructional Science, 51*(5), 687–694.

Kidron, A., & Kali, Y. (2017). Boundary breaking for interdisciplinary learning. *Research in Learning Technology, 25*, Article 1996. https://doi.org/10.25304/rlt.v25.1996

Kloetzer, L., Lorke, J., Roche, J., Golumbic, Y., Winter, S., & Jõgeva, A. (2021). Learning in citizen science. In *The science of citizen science* (Vol. 283). Springer.

Laurillard, D. (2013). *Teaching as a design science: Building pedagogical patterns for learning and technology*. Routledge.

Law, N., Ko, P., Charles, E. S., Choquette, J., Jackson, P., Zhang, C., Hernández-Leo, D., Kali, Y., Sagy, O., Apple, T., Levy, K. S., Papanikolaou, K., Zalavra, E., Sgouropoulou, C., Teo, C. L., Tan, Y. Y., Yuan, G., Topali, P., Ortega-Arranz, A., Villagrá-Sobrino, S. L., Asensio-Pérez, J. I., Martínez-Monés, A., Dimitriadis, Y., Zhang, J., Niu, X., Huang, Y., & Yang, Y. (2024). Roles of learning design tools in scaffolding teacher participation and learning in research practice partnerships. In R. Lindgren, T. I. Asino, E. A. Kyza, C. K. Looi, D. T. Keifert, & E. Suárez (Eds.), *Proceedings of the 18th International Conference of the Learning Sciences – ICLS 2024* (pp. 1902–1909) International Society of the Learning Sciences.

Lemke, J. L. (1990). *Talking science: Language, learning, and values*. Ablex Publishing Corporation.

Matuk, C., Yetman-Michaelson, L., Martin, R., Vasudevan, V., Burgas, K., Davidesco, I., Shevchenko, U., Chaloner, K., & Dikker, S. (2023). Open science in the classroom: Students designing and peer reviewing studies in human brain and behavior research. *Instructional Science, 51*(5), 793–845.

McKenney, S., & Reeves, T. (2018). *Conducting educational design research*. Routledge.

OECD. (2018). The future of education and skills: Education 2030. http://www.oecd.org/education/2030-project/about/documents/E2030%20Position%20Paper%20(05.04.2018).pdf

Pandya, R., & Dibner, K. A. (2018). *Learning through citizen science: Enhancing opportunities by design* (Consensus Study Report). National Academies Press.

Phillips, T., Porticella, N., Constas, M., & Bonney, R. (2018). A framework for articulating and measuring individual learning outcomes from participation in citizen science. *Citizen Science: Theory and Practice, 3*(2), Article 3. https://doi.org/10.5334/cstp.126

Robinson, L. D., Cawthray, J. L., West, S. E., Bonn, A., & Ansine, J. (2018). Ten principles of citizen science. In S. Hecker, M. Haklay, A. Bowser, Z. Makuch, J. Vogel, & A. Bonn (Eds.), *Citizen science: Innovation in open science, society and policy* (1st ed., pp. 27–40). UCL Press.

Roche, J., Bell, L., Galvão, C., Golumbic, Y. N., Kloetzer, L., Knoben, N., et al. (2020). Citizen science, education, and learning: Challenges and opportunities. *Frontiers in Sociology, 5*, 613814. https://doi.org/10.3389/fsoc.2020.613814

Sagy, O., Golumbic, Y. N., Ben-Horin Abramsky, H., Benichou, M., Atias, O., Manor Braham, H., … & Angel, D. (2019). Citizen science: An opportunity for learning in the networked society. In *Learning in a networked society: Spontaneous and designed technology enhanced learning communities* (pp. 97–115). Springer.

Sagy, O., Kali, Y., Baram-Tsabari, A., Tal, T., & Ben-Zvi, D. (2020). *Taking citizen science to school: A mutualistic ecology of science learning*. Paper presented at the online conference for citizen and participatory science 2020, Trieste, Italy.

SDGs. (2024, March 1). https://sdgs.un.org/goals/goal4

Stake, R. E. (2013). *Multiple case study analysis*. Guilford Press.

Tal, T., Barkae, H. S., & Alon, N. L. (2023). From school-based citizen science to transition-driven activism with the community. In J. Diamond & S. Rosenfeld (Eds.), *Amplifying informal science learning: Rethinking research, design, and engagement* (1st ed., pp. 141–154). Routledge.

TCSS. (2024, March 1). https://tcss.center

Thiel, M., Baeza Álvarez, J., Diaz, M., de Veer, D., Dittmann, S., Guevara-Torrejón, V., et al. (2023). Communication strategies in an international school citizen science program investigating marine litter. *Frontiers in Environmental Science, 11*, 1270413.

Zoellick, B., Nelson, S. J., & Schauffler, M. (2012). Participatory science and education: Bringing both views into focus. *Frontiers in Ecology and the Environment, 10*(6), 310–313.

Chapter 4
Environmental Citizen Science for Environmental Citizenship: An Educational Innovation in Biology and Environmental Education for Green Transition and Sustainable Development

Andreas Ch. Hadjichambis, Demetra Hadjichambi, and Yiannis Georgiou

4.1 The Current Environmental Crisis and the Need for Education for Environmental Citizenship

The current environmental and climate crisis experienced by the planet and humanity has reached dramatic levels resulting in various global environmental problems and challenges. Climate change, loss of biodiversity, air pollution and mass destruction from extreme events, such as wild fires and massive floods, emphasize the necessity to find alternative ways of living in harmony with nature and the environment. Non-respect of environmental rights and duties, and environmental injustice are very common phenomena in our modern world (Roncucci, 2019). The intensive use of the planet's resources brings about an inter- and intra-generational injustice; namely, an environmental injustice within a generation and across different generations (Van de Velde, 2023). Moreover, human's relationship with nature in the modern world has been significantly disturbed (Beery et al., 2023). The modern human of the western world has distanced him/herself from nature and in many occasions ignores the natural environment by living in purely anthropogenic built environments or fears it by experiencing a sense of biophobia (Olivos-Jara et al., 2020).

A. C. Hadjichambis (✉) · D. Hadjichambi
Cyprus Centre for Environmental Research and Education (CYCERE), Lemesos, Cyprus

Cyprus Ministry of Education, Sport and Youth, Kimonos & Thoukididou, Nicosia, Cyprus
e-mail: a.hadjichambis@cytanet.com.cy; d.hadjichambi@cytanet.com.cy

Y. Georgiou
Department of Communication & Internet Studies, Cyprus University of Technology, Limassol, Cyprus
e-mail: ioannis.georgiou@cut.ac.cy

D. Tsybulsky, T. Tal (eds.), *Citizen Science Contributions to Biology and Environmental Education*, Contributions from Biology Education Research,
https://doi.org/10.1007/978-3-031-99298-8_4

In this urgent situation the role of citizens as "environmental agents of change" is particularly important (Walker, 2021). Today, more than ever, the citizens need to take actions to reverse this declining state of the environment (Kranz et al., 2022) and bring about an environmental and social change, acting for the benefit of the environment and, as a natural consequence, for the benefit of humanity. Citizens should be able to identify the structural causes of environmental problems as well as to achieve a critical and active engagement and civic participation for the benefit of the environment and for achieving sustainability.

Education for Environmental Citizenship may provide an ideal venue for achieving these goals (Hadjichambis et al., 2023). Education for Environmental Citizenship as defined by the European Network for Environmental Citizenship [ENEC] (2018) is:

> The type of education which cultivates a coherent and adequate body of knowledge as well as the necessary skills, values, attitudes and competences that an environmental citizen should be equipped with in order to be able to act and participate in society as an agent of change in the private and public sphere, on a local, national and global scale, through individual and collective actions, in the direction of solving contemporary environmental problems, preventing the creation of new environmental problems, in achieving sustainability as well as developing a healthy relationship with nature. Education for Environmental Citizenship (EEC) is important to empower citizens to exercise their environmental rights and duties, as well as to identify the underlying structural causes of environmental degradation and environmental problems, develop the willingness and the competences for critical and active engagement and civic participation to address those structural causes, acting individually and collectively within democratic means and taking into account the inter- and intra-generational justice.

In this direction, Hadjichambis and Paraskeva-Hadjichambi (2020), proposed the model of Education for Environmental Citizenship (EEC). In this model, they have summarized the building blocks of EEC as follows: (a) the competences that shape the personal development of the citizen (illustrated within the green core cycle), (b) the possible actions that an environmental citizen can take in different dimensions (i.e. collective and individual), spheres (i.e. public and private), scales (i.e. local, national, global), as well as (c) the main intended environmental outcomes that could contribute to environmental and social change. The model is presented at Fig. 4.1.

Emphasis and focus need to be given in key concepts such as 'achieving civic participation and critical and active engagement' and 'action taking'. Contemporary attempts of citizen science initiatives fostering environmental citizenship in biology and environmental education should empower students with the skills or competencies necessary to take part in collective actions, participatory processes and critical and active civic engagement. This attempt is relevant with democratic processes and socio-political participation and according to Schulz et al. (2016) it includes the stages of: Decision-making, Influencing, and Community participation. Considering the classification of Shirk et al. (2012), and the attempts to address socio-cultural and socio-political theoretical frames within Citizen Science (CS) initiatives, different degrees of public participation have been identified so far, classifying the levels

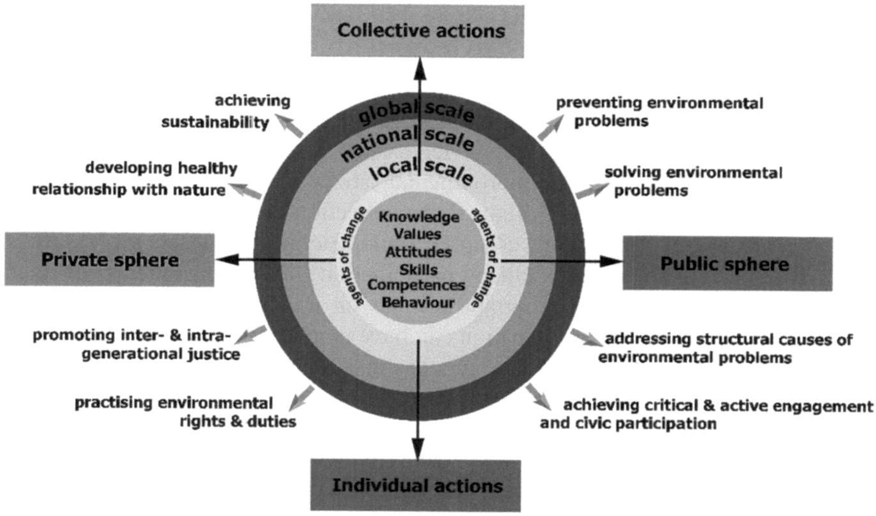

Fig. 4.1 The EEC Model. (Source: Hadjichambis & Paraskeva-Hadjichambi, 2020)

of citizens' involvement into "contractual", "contributory", "collaborative", "co-created", and "collegial". In contractual CS initiatives, the scientists conduct, for instance, an investigation requested by the communities. On the other hand, contributory and collaborative CS initiatives are designed by scientists; however, in the former case, citizens take part as "data collectors", while in the latter case the citizens also contribute to the design, data analysis, and/or dissemination of the outcomes. Co-created initiatives are co-designed by both scientists and citizens, and they are characterised by the active participation of citizens throughout the whole duration of the project. Collegial initiatives are the ones where citizens conduct an investigation independently, while expecting recognition from scientists. Despite the significance of designing collaborative and co-created initiatives, in which citizens are actively engaged, there are many CS initiatives which use contributory approaches, where citizens are simply taking the role of "passive sensors" providing the required data to the scientists (Jordan et al., 2011; Kofinas, 2009; Stewart & Sinclair, 2007).

In addition, action taking in authentic real-world contexts, is also crucial. These individual and collective actions should be taken in private and public spheres and at the same time in local, national and global scales. This is necessary if these actions need to bring environmental and social changes for the benefit of the environmental (Adamou et al., 2021; Hadjichambis & Paraskeva-Hadjichambi, 2020). Contemporary educational programs applying this philosophy (e.g., Simonova et al., 2019; Hadjichambis et al., 2022) proved to be very promising in this direction.

4.2 Citizen Science Initiatives as Springboard for the Education for Environmental Citizenship

Recent studies have indicated that environmental citizen science initiatives may support a transformative learning approach, which is a prerequisite for EEC (Bela et al., 2016; Van Wyk, 2015). Environmental citizen science initiatives, as documented in the literature, can contribute to the collection of important environmental data that are useful for solving complex socio-scientific issues (e.g., Vasiliades et al., 2021). More specifically, citizen participation (including students) in scientific research is particularly important and has an important history so far (e.g., Bonney et al., 2016). This participation can have amphidromic characteristics with benefits both to science as well as to citizens and society (Chase & Levine, 2016; Kloetzer et al., 2021). Environmental Citizen Science projects often aim to address various environmental problems, through mass data collection, which are more easily collected by citizens, and contribute to environmental protection and nature conservation. On the other hand, citizens through their participation in Environmental Citizen Science programs acquire new knowledge and skills.

However, despite these arguments, according to Jørgensen and Jørgensen (2020), at the moment, environmental citizen science initiatives place more emphasis on the scientific component (i.e. "science"), as they are designed to address merely scientific questions and ignore their "citizen" component. In addition, it is important to mention that during these projects, citizens' participation is often narrowed down to the exclusive role of "data collectors" or "passive" participants (Yadav & Darlington, 2017). According to Vasiliades et al. (2021) most of the CS initiatives reported in the literature were primarily followed the "contributory" model. In other words, these CS initiatives are designed by scientist sand citizens take part as "data sensors" collecting merely data without voice, willingness participation and action. There is therefore a need for environmental citizen science initiatives to find ways to incorporate the "citizen-centered" dimension to a greater degree. In this chapter, an attempt will be made to describe how environmental citizen science programs can focus more on "citizens" and the promotion of "environmental citizenship".

To address this issue, the EnviroCitizen (www.envirocitizen.eu) project, funded by the EU Horizon 2020 fund, investigated how environmental citizen science initiatives could be (re-)designed to support more productive forms of environmental citizenship. In this context, the Cyprus Centre for Environmental Research and Education (CYCERE), which served as a consortium partner of the EnviroCitizen project, carried out a holistic evaluation and analysis of environmental citizen science initiatives through undertaking a sequence of systematic reviews. Adopting a meta-synthesis approach, these systematic reviews were finally compiled resulting in the development of Science for Environmental Citizenship (CS4EC) framework, to provide guidance on how environmental citizen science initiatives can be (re-) designed to foster environmental citizenship. In the rest of this chapter, we will discuss and further elaborate on the findings of these systematic reviews, concluding with in an overview of the CS4EC framework.

4.3 Citizens' Participation Aspects in Environmental Citizen Science Projects

As part of the first systematic review conducted by CYCERE, Vasiliades et al. (2021) found that environmental CS initiatives have often merely focus on the achievement of their scientific goals and outcomes (science-centered), rather than on empowering and transforming the participants into 'environmental citizens'(citizen-centered). In addition, a main finding was that the reviewed environmental CS initiatives were mostly focused on recruiting adult populations rather than students, and as such they were less oriented towards educating the participants.

More specifically, analyzing the participatory aspects of a total of 119 environmental citizen science initiatives applied in the last two decades, it was found that most of the participants were adults (71%); as can be seen in Fig. 4.2, only 29% of CS initiatives recruited participants under the age of 18. On the other hand, no gender discrimination was identified, as both males and females were found to participate in the vast majority (97%) of CS initiatives. Regarding the education of the participants, it was found that 61% of the participants, were at least college degree holders indicating well educated participants. Furthermore, most of the CS initiatives reported in were locally oriented (48%) focusing on local communities, with a measurable quantity taking also place at national level (39%). However, only a limited number of citizen science initiatives expanded more broadly at a continental and international level (6% and 7%, respectively). Finally, the coordination of the environmental citizen science initiatives reviewed were attributed to Academic institutions (63%).

As already mentioned, among the reviewed environmental citizen science initiatives only 16% of them had "education" as overarching aim since more of them aiming in nature conservation and environmental awareness. In more details, "understanding of scientific content and knowledge", and "engaging in scientific practices", were the main science related learning goals of the reviewed environmental CS initiatives.

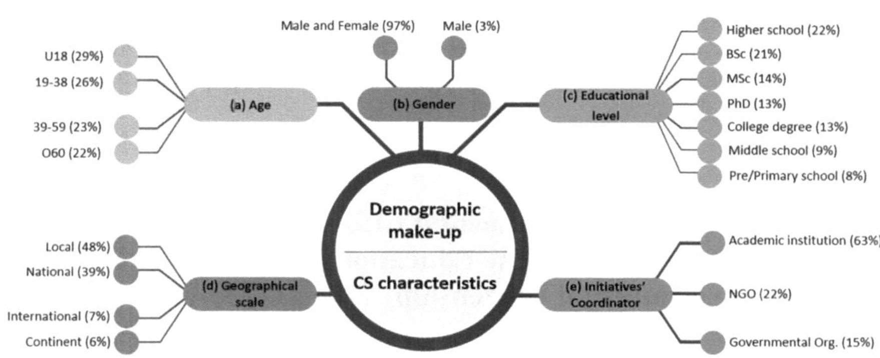

Fig. 4.2 Participation aspect. (Source: Vasiliades et al., 2021)

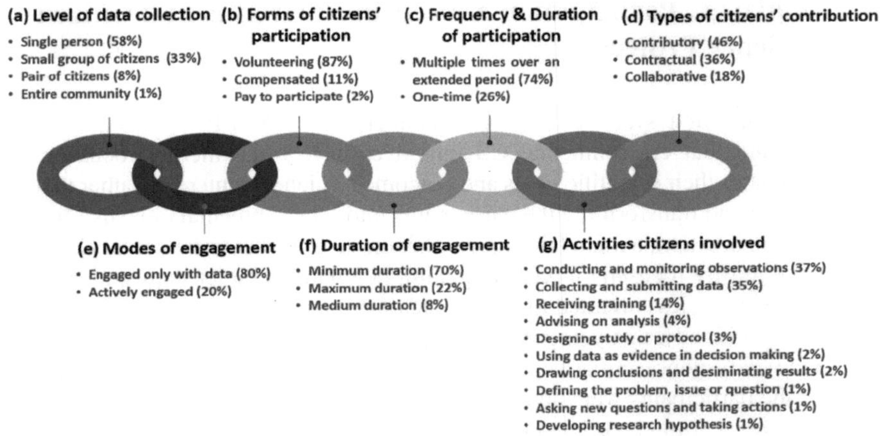

Fig. 4.3 (a) Level of data collection (n = 68), (b) Forms of citizens' participation (n = 74), (c) Frequency & duration (n = 68) of citizens' participation, (d) Types of citizens' contribution, (n = 65), (e) Modes of engagement (n = 57), (f) Duration of engagement (n = 50), and (g) Citizens' activities (n = 63) in CS initiatives. (Source: Vasiliades et al., 2021)

In the same review, it was also found that most of the environmental CS initiatives were contributory (46%) or contractual (36%) with only 18% to characterised as collaborative (Fig. 4.3). In addition, citizens' participation was mainly grounded on volunteering basis (87%), and was characterized by single-person on data collection (58%) as well as by a minimum duration of engagement (70%). Importantly, in their vast majority (80%), citizens were engaged only with data confirming their science-centered character.

Finally, during this systematic literature review, it was found that most of the environmental CS initiatives targeted the "individual" (66%) over the "collective" (34%) actions of environmental citizenship. Also, mostly included "private" sphere actions (70%) rather than actions in the "public" sphere (30%). The main facilitators of the environmental citizen science studied were: Contribute to Science (23%), Learning Gains (21%) and Connecting with Nature (21%). As far as Constraints, the main constraints identified were: some Design matters of CS and platforms (31%), the Time requirements (15%), the need of Special equipment (8%) and the need of Funding/cost (7%).

4.4 How Current Environmental Citizen Science Initiatives Contribute Towards the Education for Environmental Citizenship

A subsequent question in relation to the current status of citizen science, was to identify (a) whether participation in environmental citizen science initiatives contributes to the development of citizens' environmental citizenship competences, actions and outcomes as well as (b) the pedagogical aspects contributing to EC. Toward this end, the CYCERE research team carried out two additional systematic literature reviews published by Adamou et al. (2021) as well as by Hadjichambi et al. (2023), exploring the empirical research published during the last two decades (2000–2020). What follows is an overview of the main findings derived from the aforementioned reviews.

4.4.1 Impact on EC Competences

According to the findings derived from the systematic review of Adamou et al. (2021) the majority of the reviewed environmental CS initiatives mainly strengthened citizens' knowledge and skills over pro-environmental attitudes, values and behaviors. More specifically, the reviewed environmental CS initiatives mainly contributed to the enhancement of the citizens' knowledge (n = 25 CS initiatives, 80.5%), and skills (n = 27 CS initiatives, 87.1%), in order to be able to collect reliable datasets and to a lesser degree to the enhancement of attitudes (n = 19 CS initiatives, 61.3%), values (n = 11 CS initiatives, 35.5%), and behaviors (n = 8 CS initiatives, 25.8%).

In more detail, regarding Knowledge the reviewed CS initiatives promoted the enhancement of Environmental Systems Knowledge (ESK) (n = 19 CS initiatives, 61.3%), Self-Effectiveness Knowledge (SEK) (n = 11 CS initiatives, 35.5%), and Action-Related Knowledge (ARK) (n = 11 CS initiatives, 35.5%). As far as it concerns the EC skills, they proved to enhance Collaboration and Social Interaction Skills (CSIS), namely to the ability of individuals to effectively engage with and co-work with others in common or public interest, in relation to environmental issues (n = 16 CS initiatives, 51.6%), Evidence-Based Thinking Skills (EBTS) (n = 16 CS initiatives, 51.6%), as well as Interacting with Media Skills (IMS) (n = 13 CS initiatives, 41.9%).

Even though to a lesser degree, the reviewed environmental CS initiatives empowered also pro-environmental attitudes such as the Willingness for Collective Environmental Actions (WCEA) (n = 14 CS initiatives, 45.2%), Willingness to Act in Society as Agent of Change (WASAC) (n = 10 CS initiatives, 32.3%), Willingness to Eliminate New Environmental Problems (WENEP) (n = 8 CS initiatives, 25.8%) or Willingness for Networking to solve environmental problems (WNSEP) at the local, national and global scale (n = 4 CS initiatives, 12.9%). As far as it concerns values, we found that the reviewed CS initiatives mostly contributed to the development of Biospheric Values (BV) (n = 16 CS initiatives, 51.6%). Finally, regarding Behaviors, they had a positive effect in terms of Private Sphere Behaviors (PSB) (n = 6 CS initiatives, 19.4%).

4.4.2 Impact on EC Actions and Outcomes

Regarding EC actions, according to the systematic review of Adamou et al. (2021), the reviewed environmental CS initiatives were found to lead to the empowerment of citizen's Individual Actions (IA) (n = 19 CS initiatives, 59.4%), Private Sphere Actions (PrSA) (n = 11 CS initiatives, 35.5%) as well as Local Scale Actions (LSA) (n = 11 CS initiatives, 35.5%) and National Scale Actions (NSA) (n = 10 CS initiatives, 32.3%).In addition, regarding EC outcomes, the reviewed environmental CS initiatives, proved to contribute mainly to the Development of Healthy Relationship with Nature (HRN) (n = 15, 48.4%), to the Prevention of New Environmental Problems (PNEP) (n = 14, 45.2%) as well as to the Solution of Environmental Problems (SEP) (n = 13, 41.9%) and in lesser level to the Achievement of Sustainability (AS) (n = 6, 19.4%).

4.4.3 Keystone and Peripheral Components in Environmental CS Initiatives

Based on the correlation analysis and a k-means cluster analysis carried out it was found that environmental CS initiatives (Fig. 4.4) have seven Keystone Components (KCs, n = 7 EC components) which indicated a higher number of connections (x = 7, SD = 1.73), in comparison with the Peripheral Components group (PCs, n = 41 Components), which indicated a lower number of connections. The seven KCs identified are Willingness to Act in Society as Agents of Change (WASAC) (n = 9 connections), Solution of Environmental Problems (SEP) (n = 9 connections), Political Systems Knowledge (PSK), (n = 8 connections), System Thinking Skills (STS) (n = 7 connections), Constructive Participation Skills (CPS) (n = 6 connections), Problem-Solving Skills (PSS) (n = 5 connections), and Local Scale Actions (LAS) (n = 5 connections). These EC components considered to be the most salient components if the environmental CS initiatives have to contribute to citizens' environmental citizenship. These findings may guide the future design of environmental CS initiatives, as according to Jørgensen and Jørgensen (2020), involvement in environmental CS activities can cultivate environmental citizenship, if CS initiatives are intentionally designed to do so.

4.4.4 The CS Pedagogical Aspects Contributing to EC

In the systematic review study of Hadjichambi et al. (2023), which studied K-12 environmental CS initiatives that had educational goals, it was confirmed that the reviewed educational environmental CS initiatives had positive effects mainly on students' knowledge and skills. Next, regarding the EC outcomes we found that the

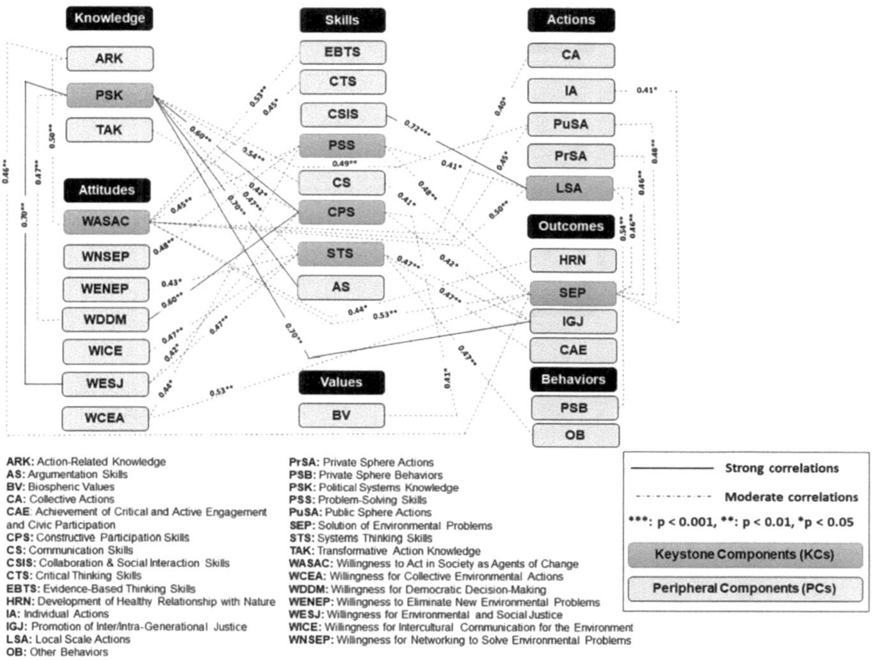

Fig. 4.4 Correlations between the keystone components with the other EC components. (Source: Adamou et al., 2021)

reviewed educational CS initiatives had positive impacts on developing healthy relationship with nature and in achieving a solution of environmental problems. In the same study it was also found that most of the reviewed educational CS initiatives (a) were contributory, (b) comprised of small groups of students, (c) extended over multiple periods of time, and (d) exclusively engaged students with data collection, given that (e) they mostly required students to gather and submit data/samples. Then, regarding their underpinning pedagogical components, it was documented that the reviewed CS initiatives mainly (a) sought to improve students' knowledge on a specific scientific topic, (b) were situated in non-formal educational contexts, (c) were scaffolded via the use of educational resources, as well as by the participation of expert mentors, (d) were structured around students' contribution in the undertaken tasks, and (e) most of them, adopted the inquiry-based approach.

As in the previous review study by Adamou et al. (2021), in this study the Keystone Components (KC) were also identified after conducting a correlation analysis and a k-means cluster analysis. Based on these analyses, five KCs were identified (Fig. 4.5) as follows: Collective Actions (CA) ($n = 7$ connections), Civic Participation and Critical Active Engagement (CPCAE) ($n = 6$ connections), Communicate Findings (CF) ($n = 4$ connections), Planning Actions (PA) ($n = 5$ connections) and Public Sphere Actions (PuSA) ($n = 4$ connections).

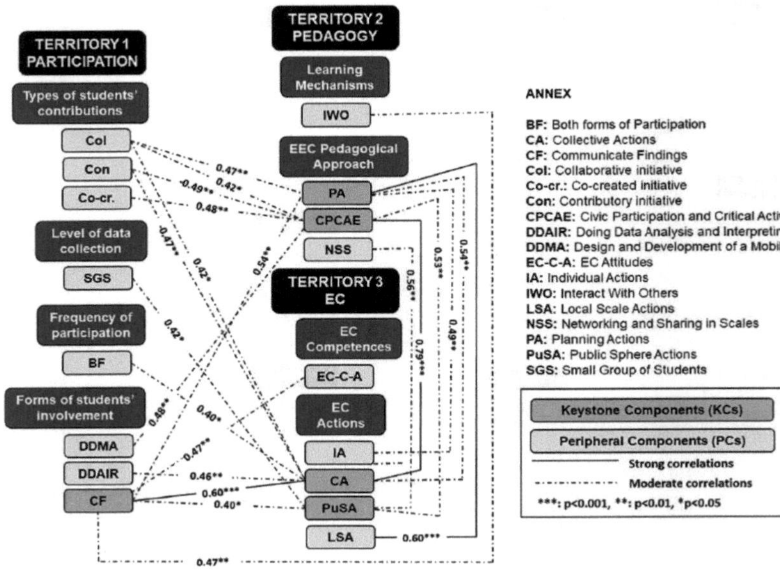

Fig. 4.5 Correlations between the keystone components with the peripheral components. (Source: Hadjichambi et al., 2023)

The results of this study are considerably useful for the design of future educational environmental CS projects and may contribute to the transformation of students into young environmental citizens capable to act as agents of change in order to achieve socio-environmental justice. It is also important, for future educational environmental CS initiatives to be designed in a way that will also include Collective Actions (CA) and especially actions in the Public Sphere (PuSA). Greater emphasis should also be given on outcomes such as promoting Inter/Intra-Generational Justice (IGJ), on the empowerment of practicing Environmental Rights and Duties (ERD), on addressing the Structural Causes of Environmental Problems (SCEP), on achieving Critical and Active Engagement and Civic Participation (CAE), on Achieving Sustainability (AS), as well as on Preventing the creation of New Environmental Problems (PNEP).

4.5 How Education in Environmental Citizen Science Could Contribute Better to Environmental Citizenship: The CS4EC Theoretical Framework

The ultimate goal of the meta-synthesis carried out by CYCERE research team was to synthesize evidence from the available systematic reviews, seeking to extract a set of design principles for the development of an integrated theoretical framework enabling Environmental Citizenship (EC) in environmental CS initiatives. The Citizen Science for Environmental Citizenship (CS4EC) framework (Hadjichambis et al., 2023) includes a transformative potential and fosters participatory learning aiming to result in environmental CS initiatives capable of empowering and transforming citizens into "environmental change agents".

The Citizen Science for Environmental Citizenship (CS4EC) framework sets out 15 key principles for designing an environmental citizen science program on the basis of three sequential phases: Citizens (including students) as "co-designers", citizens as "scientists" and citizens "agents of change". Figure 4.6 presents the procedural model of the CS4EC theoretical framework as well as they key design principles underpinning each phase.

The first phase called "citizens as co-designers" involves the participation of citizens (including students) in the co-creation of a citizen science program to study an environmental problem. Therefore, the starting point includes the principles of inclusiveness with the participation of both adults (scientists, teachers, etc.) and students, as well as the participation of citizens from marginalized groups and groups with disabilities. For the co-creation of the initiative, the common criteria, the research questions, and the objectives of the program are collectively defined by all the involved stakeholders.

The second phase is the phase where citizens (including students) act as scientists. During this phase, following an inquiry-based learning process, the CS participants interact in an authentic educational context, collect data in a collaborative working framework, they adopt active and social learning in the context of an extensive collaboration, using multiple educational tools and resources, and finally, they also communicate their results thus achieving a multiplier effect. In this phase, special importance should be also given to Keystone Components such as political systems knowledge, systems thinking skills, constructive participation skills and problem-solving skills.

The third and final phase of the CS4EC model, citizens (including students) move beyond inquiry and act as "agents of change". In this phase they plan both

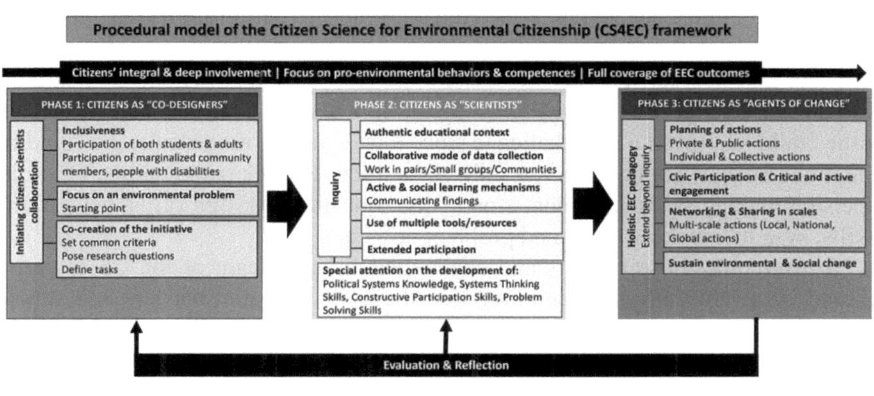

Fig. 4.6 The procedural Model of the CS4EC framework. (Source: Hadjichambis et al., 2023)

individual and collective actions in the private and public spheres. They engage in extensive civic participation and critical and active engagement. They also carry out extensive networking and sharing in different scales (local, national, global). Finally, they proceed in additional actions to sustain environmental and social change.

4.6 Challenges and Potential Barriers to Implementing the CS4EC Framework

In this section, challenges and potential barriers to implementing the CS4EC framework are discussed since such a discussion would strengthen model's practical applicability. In addition, discussion on strategies to overcome these challenges were elaborated.

The CS4EC model is a holistic theoretical model based on a meta-synthesis of extensive review of empirical studies. For this reason, it may face some challenges and some possible barriers for its application by the practitioners. One of its main characteristics is its holistic nature. It brings together and combines elements from both citizen science, environmental science and pedagogy. In addition, it combines the principle of co-creation, participatory and democratic education, authentic social learning, intergenerational learning including students and adults, change orientation and inclusiveness approach. This can be a great advantage from its holistic side but on the other hand it can be a great challenge to put into practice.

At the same time, it involves the participation of academics, scientists and researchers who are experts in the authentic environmental problem to be studied, teachers and students, social partners and stakeholders, IT specialists, volunteers and amateurs, government departments, lawyers and parliamentary representatives as well as businessmen and local governments. All these versatile groups involved may have different views, motivations and goals and therefore it can be a big challenge to work together and reach a consensus. A consensus should be reached on a range of issues such as the CS initiatives' research questions, design principles, educational approaches and multiple educational tools, actions to be taken at local, national and global scales.

Other possible barriers include stakeholder resistance and logistical constraints as well as the range of necessary skills by the core team and participants. This latter issue could be proved a real challenge since a wide range of needed skills could be identified such as communication skills, constructive participation skills, IT skills, decision making skills, argumentation skills, educational design skills, networking skills, systems thinking skills and problem-solving skills.

To overcome those barriers and challenges the core team of the CS initiative should be aware from the beginning that these are potential barriers and should define a clear strategy how to overcome these barriers from the beginning of the co-design of the environmental CS initiative. For example, a clear strategy is essential to incorporate the critical mass of the necessary expertise and the needed skills.

Another example includes a strategy to solve and overcome disagreements and conflicts. In addition, "previous experience" and "multiple roles" could also be proved crucial elements for the successful implementation of the CS4EC theoretical framework and its procedural model. For example, a multiple role of experienced teachers who have also a background in research and in educational co-design could be beneficial for the CS initiative. Another example of multiple roles could be an academic and a researcher who has also experience in IT design and in educational co-design. Such overlaps in expertise and in multiple roles could be proved very beneficial for the implementation of the CS4EC as they minimize the "distance" between different expertise and different perspectives. Finally, elements of the Education for Environmental Citizenship pedagogical approach (Hadjichambis & Paraskeva-Hadjichambi, 2020) could also be beneficial in the implementation of the CS4EC as it incorporates lot of the elements of the CS4EC and successful applications of this pedagogical approach are already published (Hadjichambis et al., 2022).

4.7 Contribution and New Connections

This chapter delves into the transformative potential of Environmental Citizen Science (ECS) initiatives in cultivating environmental citizenship. Amidst escalating global environmental crises, the chapter argues for a paradigm shift from "science-centric" to "citizen-centric" approaches in ECS projects. Through a comprehensive analysis, it introduces the Citizen Science for Environmental Citizenship (CS4EC) framework, aiming to foster environmental citizenship by involving citizens, including students, in co-designing, data collection and decision-making processes, concerning socio-environmental issues. This chapter is emphasizing the role of citizen science in citizen science initiatives in fostering environmental citizenship, especially among school participants. This presents a clear step forward the field of citizen science as well as in the field of biology and environmental education.

The chapter is meticulously structured, facilitating an accessible yet profound exploration of the intersection between environmental citizen science and environmental citizenship. It commences with a compelling narrative on the urgency of addressing the environmental crisis through innovative educational approaches in the field of biology and environmental education, setting a solid foundation for the subsequent discussion. The progression from identifying the gap in current ECS initiatives is enriched by relevant case studies and systematic reviews.

This chapter significantly advances the discourse on environmental education and citizen science by proposing a well-conceived framework that shifts the focus towards a more inclusive, participatory and action-oriented approach. Its contribution lies not only in critiquing the prevailing science-centered models but also in offering a practical, theoretically underpinned pathway towards empowering citizens as agents of environmental change. This is particularly relevant for educators, policy-makers and researchers seeking to enhance the efficacy and impact of ECS initiatives.

In addition, the chapter in its comprehensive synthesis of literature and empirical studies, culminating in the CS4EC framework which gives emphasis on inclusivity, co-creation and the transformative potential of ECS initiatives, provides novel lens and new connections through which to view and enhance environmental citizenship. Additionally, the chapter is making complex theoretical concepts accessible, accompanied by clear illustrations and examples that illuminate the proposed ideas. Finally, the chapter is discussing challenges and potential barriers of implementing the CS4EC framework and facilitates possible strategies to overcome those challenges for the benefit of biology and environmental education as well as for green transition (Hadjichambis, 2022).

4.8 Conclusion

In conclusion, environmental CS initiatives have a huge potential in biological and environmental education as a promising channel for achieving green transition and sustainable development. However, they should escape from their current practice, as in many cases they limit citizens (including students) to the collection of scientific data. This chapter provides the necessary insights for environmental CS initiatives to take a more holistic approach grounded on the Education for Environmental Citizenship, as a venue through which citizens can undertake actions in different scales (local, national, global) to achieve environmental citizenship. The Citizen Science for Environmental Citizenship (CS4EC) framework, as presented in this chapter, incorporates all those features that will empower environmental citizens to undertake environmental and social change as well as to achieve sustainability.

We also consider the CS4EC framework, as a powerful framework for re-shaping and supporting teachers' understanding of environmental citizenship. Undoubtedly, teachers are considered as the main driving force in fostering students' environmental citizenship. However, a recent systematic review by Georgiou et al. (2021) has indicated that teachers manifest a relatively decreased understanding of environmental citizenship and this negatively affects their teaching practices. In this context, we argue that the integration of the CS4EC framework in pre-service teachers' studies as well as in subsequent professional development programmes may improve teachers' perceptions of what environmental citizenship is, what it entails and how it can be fostered, through the appropriate design of environmental CS initiatives. The CS4EC framework could also allow teachers to join forces with scientists through school-universities-research centers partnerships for co-designing environmental CS programmes geared towards empowering students' environmental citizenship. In turn, CS programmes with a clear focus on environmental citizenship could be considered as an educational innovation for green transition and sustainable development.

References

Adamou, A., Georgiou, Y., Paraskeva-Hadjichambi, D., & Hadjichambis, A. C. (2021). Environmental citizen science initiatives as a springboard towards the education for environmental citizenship: A systematic literature review of empirical research. *Sustainability, 13*(24), 13692.

Beery, T., Stahl Olafsson, A., Gentin, S., Maurer, M., Stålhammar, S., Albert, C., Bieling, C., Buijs, A., Fagerholm, N., Garcia-Martin, M., Plieninger, T., & Raymond, C. (2023). Disconnection from nature: Expanding our understanding of human–nature relations. *People and Nature, 5*(2), 470–488.

Bela, G., Peltola, T., Young, J. C., Balázs, B., Arpin, I., Pataki, G., ... Bonn, A. (2016). Learning and the transformative potential of citizen science. *Conservation Biology, 30*(5), 990–999. https://doi.org/10.1111/cobi.12762

Bonney, R., Phillips, T. B., Ballard, H. L., & Enck, J. W. (2016). Can citizen science enhance public understanding of science? *Public Understanding of Science, 25*(1), 2–16.

Chase, S. K., & Levine, A. (2016). A framework for evaluating and designing citizen science programs for natural resources monitoring. *Conservation Biology, 30*(3), 456–466. https://doi.org/10.1111/cobi.12697

European Network for Environmental Citizenship – ENEC. (2018). *Defining "education for environmental citizenship"*. Retrieved from http://enec-cost.eu/our-approach/education-for-environmental-citizenship/

Georgiou, Y., Hadjichambis, A. C., & Hadjichambi, D. (2021). Teachers' perceptions on environmental citizenship: A systematic review of the literature. *Sustainability, 13*(5), 2622.

Hadjichambi, D., Hadjichambis, A. C., Adamou, A., & Georgiou, Y. (2023). A systematic literature review of K-12 environmental citizen science (CS) initiatives: Unveiling the CS pedagogical and participatory aspects contributing to students' environmental citizenship. *Educational Research Review, 39*, 100525.

Hadjichambis, A. C. (2022). European green deal and environmental citizenship: Two interrelated concepts. *Environmental Sciences Proceedings, 14*(1), 3.

Hadjichambis, A. C., & Paraskeva-Hadjichambi, D. (2020). Education for environmental citizenship: The pedagogical approach. In A. C. Hadjichambis, P. Reis, D. Paraskeva-Hadjichambi, et al. (Eds.), *Conceptualizing environmental citizenship for 21st century education* (pp. 262–292). Springer. ISBN 978-3-030-20249-1.

Hadjichambis, A. C., Paraskeva-Hadjichambi, D., & Georgiou, Y. (2022). Evaluating a novel learning intervention grounded in the education for environmental citizenship pedagogical approach: A case study from Cyprus. *Sustainability, 14*(3), 1398.

Hadjichambis, A. C., Paraskeva-Hadjichambi, D., Georgiou, Y., & Adamou, A. (2023). How can we transform citizens into 'environmental agents of change'? Towards the citizen science for environmental citizenship (CS4EC) theoretical framework based on a meta-synthesis approach. *International Journal of Science Education, Part B*, 1–21.

Jordan, R. C., Gray, S. A., Howe, D. V., Brooks, W. R., & Ehrenfeld, J. G. (2011). Knowledge gain and behavioral change in citizen-science programs. *Conservation Biology, 25*(6), 1148–1154.

Jørgensen, F. A., & Jørgensen, D. (2020). Citizen science for environmental citizenship. *Conservation Biology, 35*(4), 1344.

Kloetzer, L., Lorke, J., Roche, J., Golumbic, Y., Winter, S., & Jõgeva, A. (2021). Learning in citizen science. In K. Vohland, A. Land-Zandstra, L. Ceccaroni, R. Lemmens, J. Perelló, M. Ponti, & K. Wagenknecht (Eds.), *The science of citizen science* (pp. 283–308). Springer.

Kofinas, G. P. (2009). Adaptive co-management in social-ecological governance. In *Principles of ecosystem stewardship: Resilience-based natural resource management in a changing world* (pp. 77–101). Springer.

Kranz, J., Schwichow, M., Breitenmoser, P., & Niebert, K. (2022). The (Un) political perspective on climate change in education—A systematic review. *Sustainability, 14*(7), 4194.

Olivos-Jara, P., Segura-Fernández, R., Rubio-Pérez, C., & Felipe-García, B. (2020). Biophilia and biophobia as emotional attribution to nature in children of 5 years old. *Frontiers in Psychology, 11*, 511.

Roncucci, R. (2019). *Rights of nature and the pursuit of environmental justice in the Atrato case*. Wageningen University & Research.

Schulz, W., Ainley, J., Fraillon, J., Losito, B., & Agrusti, G. (2016). *IEA international civic and citizenship education study 2016 assessment framework*. IEA. http://research.acer.edu.au/civics/24

Shirk, J. L., Ballard, H. L., Wilderman, C. C., Phillips, T., Wiggins, A., Jordan, R., McCallie, E., Minarchek, M., Lewenstein, B. V., Krasny, M. E., & Bonney, R. (2012). Public participation in scientific research: A framework for deliberate design. *Ecology and Society, 17*(2), 29.

Simonova, P., Cincera, J., Kroufek, R., Krepelkova, S., & Hadjichambis, A. (2019). Active citizens: Evaluation of a community-based education program. *Sustainability, 11*(3), 663.

Stewart, J. M., & Sinclair, A. J. (2007). Meaningful public participation in environmental assessment: Perspectives from Canadian participants, proponents, and government. *Journal of Environmental Assessment Policy and Management, 9*(02), 161–183.

Van de Velde, C. (2023). 'What have you done to our world?': The rise of a global generational voice. *International Sociology, 38*(4), 431–457. https://doi.org/10.1177/02685809231180880

Van Wyk, S. M. (2015). *Environmental citizenship in citizen science: A case study of a volunteer toad conservation group in Noordhoek, South Africa* [Doctoral dissertation, Rhodes University].

Vasiliades, M. A., Hadjichambis, A. C., Paraskeva-Hadjichambi, D., Adamou, A., & Georgiou, Y. (2021). A systematic literature review on the participation aspects of environmental and nature-based citizen science initiatives. *Sustainability, 13*(13), 7457.

Walker, C. (2021). 'Generation Z' and 'second generation': An agenda for learning from cross-cultural negotiations of the climate crisis in the lives of second-generation immigrants. *Children's Geographies, 19*(3), 267–274.

Yadav, P., & Darlington, J. (2017). Conceptual frameworks for building online citizen science projects. *arXiv preprint arXiv:1704.05084*.

Part II
Incorporating Citizen Science in K-12 Biology and Environmental Education: Opportunities and Challenges

Chapter 5
Engaging Students with Citizen Science Data: Gateway to New Scientific Frontiers

Michal Dvir and Dina Tsybulsky

5.1 Introduction

The relatively new scientific genre of citizen science has opened the door not only for the generation of new, previously unattainable, scientific insight (Kelling et al., 2015) but also for new educational horizons (Schuttler et al., 2019). The novel pedagogical affordances of engaging students in citizen science stem from the genre's unique characteristics. One of these is the nature of the extensive data that are generated, quite often with the assistance of the participating public (or students). Beyond being large in scale, citizen science project-generated data have unique complexities. These complexities can be seen as hurdles for novices' meaningful engagement with citizen science data. However, if the novices are fruitfully supported to overcome these challenges, the same complexities can serve as a bountiful spring of novel pedagogical opportunities. These include innovative ways to introduce novices to new (scientific) data practices, thus addressing a relatively new educational demand brought on by the information age. While traditional empirical scientific methods focus on generating data to test a pre-formulated hypothesis (Bauer, 1994), new data-driven approaches that often utilize pre-existing data (Cao, 2018) are becoming more widespread. Engaging students with citizen science data can serve as a gateway to newer data-driven scientific practices. However, this shift requires

M. Dvir (✉)
Faculty of Education in Science and Technology, Technion Israel Institute of Technology, Haifa, Israel

Faculty of Education, University of Haifa, Haifa, Israel

D. Tsybulsky
Faculty of Education in Science and Technology, Technion Israel Institute of Technology, Haifa, Israel
e-mail: dinatsy@technion.ac.il

D. Tsybulsky, T. Tal (eds.), *Citizen Science Contributions to Biology and Environmental Education*, Contributions from Biology Education Research,
https://doi.org/10.1007/978-3-031-99298-8_5

not only an understanding of the practical complexities of working with large-scale and diverse datasets but also a deeper reflection on the epistemological nature of the data. What do data represent? Are data an objective record of reality (positivist view), or are they shaped by the contexts in which they are collected and interpreted (critical realist view)? Citizen science data, in particular, challenge simplistic notions of data as neutral facts, as they are often influenced by varying collection methods, participant motivations, and diverse data sources. Recognizing these nuances is essential for designing pedagogical interventions that help novices navigate both the technical and conceptual challenges of engaging with such data. The goal of this chapter is to offer an initial framework that characterizes the unique features of engaging with citizen science data, revealing the potential challenges that novices might encounter. We then suggest a variety of activities and approaches that can inform educational designs to support novices' engagement with different aspects of citizen science data to maximize the pedagogical potential of this engagement. We conclude with an illustrative example of an educational design that implemented these suggestions.

5.2 Three Dimensions of Engaging with Citizen Science Data: Scientific, Secondary, and Classical

5.2.1 The Scientific Dimension

Data has an instrumental role in any scientific research, as the goal of scientific inquiry is often claimed to be coordinating between theory and data (Chalmers, 2013). Data is at the heart of the creation and refinement of scientific insight and distinguishes what can be considered as science from other disciplines (Popper, 1963). The relation between theory and data is described as bidirectional: existing scientific theories provide predictions of what data can be generated in a wide variety of conditions. By doing so, with the goal of examining their validity, scientific theories inform what data should be collected and in what ways. However, the data that were generated can confirm but also refute the theory that inspired their collection, thus either extending or challenging it and inspiring new scientific theories (Gasparatou, 2017).

Thus, engaging with citizen science data, as is the case with other scientific data, calls for engaging in this bidirectional coordination between theory and data. Specifically, this implies that for students to meaningfully engage with these data, they need to become familiar with pre-existing theories that inspired the data collection they assist in and come to appreciate why and how the data collection protocol was informed by them. Having completed their citizen science data investigation students should also be encouraged to relate their findings back to these theories and consider in what ways their data have potentially informed those pre-existing theories. While these are important aspects of students' engagement with citizen science data, they seem to be mostly relevant to the onset and the conclusion of this engagement and reveal little about the uniqueness of citizen science data.

5.2.2 Adding the Classical Data Dimension

To further characterize additional aspects of the engagement with citizen science data, we also must consider the fact that these are, well, data. We thus turn to the abundance of research that can be found in Statistics Education literature on novices' engagement with "classical data" (i.e., relatively small amount of cases, relatively few attributes for each case, and generated meticulously to examine a specific a-priori research question, e.g., Ben-Zvi, 2004; Biehler et al., 2018; Konold & Higgins, 2003), and on how to support their engagement (e.g., McClain et al., 2000). In many cases, engagement with data is described as a cyclic activity that follows a data investigation cycle or process (e.g., Wild & Pfannkuch, 1999). For example, Friel and Bright (1998) offer a four staged-depiction of the data investigation cycle that consists of (1) asking a question; (2) collecting data; (3) analyzing the data; and (4) forming and communicating conclusions.

Konold and Higgins (2003) elaborated on the reasoning and activities associated with each of these stages. They claim that for one to "ask a question," one must go through a process they refer to as *turning observation into data*. This process first necessitates *differentiating between* the observed *event and the data*, or acknowledging that the data are mere representations of only some of the features of the investigated phenomenon. This distinction is key in *formulating a statistical question* (rather than an interesting question, for example) and choosing proper *sampling* methods to elevate the data's representativeness. Konold and Higgins (2003) also add to the "collect data" stage the consideration of how to document the collected data and how to then *organize and display* the data in preparing for their subsequent analysis. They also elaborated on the third "analyze the data" stage, indicating that it includes both the need to *describe* the data and to *interpret* it. Konold and Higgins (2003) also added that for one to fruitfully form and communicate a conclusion, one must *relate the data back to the* observed *event* (e.g., associate the real-world meaning with the numbers the data consist of or form generalizations beyond the data at hand).

Returning now to our exploration of the unique nature of citizen science data, engaging with them will likely include some if not all of these aspects, as well as the scientific aspects of *understanding how theory informed the data* collection (as part of the *turning observation into data* stage, and *considering how the data informed* these or new *theories*). Table 5.1 offers an integrative summary of these four data investigation stages, the aspects of engaging with classical data, and scientific data associated with each aspect.

Table 5.1 Integrative summary of aspects of engaging with classical data and scientific data within each data investigation stage

Stage of the classical data investigation	Aspects of engaging with classical data	Aspects of engaging with the scientific data
Turning observation into data	Differentiating between event and data	Understanding how the theory informed the data
	Formulating a statistical question	
	Choosing sampling method	
Generating and preparing data	Collecting data	
	Documenting and organizing data	
	Displaying data	
Analyze the data	Describing the data	
	Interpreting the data	
Form and communicate conclusions	Relating the data back to the event	Considering how the data informed theory

5.2.3 Adding the Secondary Data Dimension

While the scientific context of engaging with citizen science data explains some of the data's unique features, it still does not differentiate between this engagement and the engagement in other scientific data investigations (e.g., Kanari & Millar, 2004). To deepen the discussion on its unique features, we must acknowledge the unique features of data collected in the context of citizen science, or specifically what characterizes the data that are generated from engaging the public in a citizen science project. These features include: the integration of many more people, rather than just a few scientists, resulting in large amounts of data; the data was collected by many different individuals, thus would likely result in inconsistencies in some of the data documentation; and data collected by individuals with varying levels of expertise and training, sometimes leading to errors, partial records, or challenges in data validation (Kjelvik & Schultheis, 2019). Additionally, some citizen science data is opportunistic—collected without a consistent or comparable protocol—although certain projects do implement rigorous methodological frameworks. Many of these features have been central to research in the field of Data Science education, which has made significant progress in understanding how to support novices in working with large-scale or non-traditional datasets (e.g., Gould, 2021; Rubin, 2021).

Particularly relevant are works that examined novices' engagement with 'already collected data' (Rubin, 2021), often referred to as secondary data analysis (Gould, 2021). Secondary data analysis is the examination of data that were previously collected by others, often for a different purpose than their original use. In the case of citizen science data, although students (or the assisting public) often participate in the data collection, as opposed to secondary data analysis, many of the aspects of the data generation process are pre-chosen by the citizen science scientists and are unexplained or unspecified. For example, while in classical data analysis, one forms an a-priori statistical question that guides the data collection (Bargagliotti et al., 2020), students often first participate in the creation of the data that they help generate and only later have the opportunity to ponder which questions the data can address. We, therefore, refer to citizen science data as semi-secondary, meaning that despite the students' participation in its collection, engaging with it requires some aspects of engaging with secondary data.

One distinction between engaging with secondary and classical data is that the initial stage does not focus on transforming observation into data but rather on *Transforming data into observation*. Engaging with already collected data requires one to *imagine the observed event* that the data were meant to represent and potential *additional aspects* that were not represented by the data. As in classical data analysis, the analysis should be guided by a statistical question but one that focuses on *what the data* at hand *can answer* (rather than what is interesting about the event that we can collect data to examine). An additional distinction related to the second stage, aside from the fact that the data do not warrant "generation" as they already exist, relates to more *extended preparations*. These include more extensive *cleaning* (Chu et al., 2016) actions (especially in the case of citizen science data that was generated by multiple untrained participants) and transforming the *structure* of the data to address the new purposes of the data analysis (i.e., 'data moves' such as creating new attributes, Erickson et al., 2019). One final distinction relates to the third stage of analyzing the data, in which not only the data warrant *interpretations* but also aspects of the *metadata*, or the process through which the data were generated, which is often not transparent. This includes engaging in interpreting the meaning of an *attribute's* pre-chosen *name*, the meaning of a *quantitative* entry in the data, or identifying the units used (Bargagliotti et al., 2020). All of these are also relevant to citizen science data, where the scientists often pre-chose these features in ways that are not necessarily specified to the participating public.

Figure 5.1 provides an integrative summary of these additional aspects associated with secondary data analysis that also characterizes aspects of engaging with citizen science semi-secondary data. Putting these alongside the aspects related to classical and scientific data provides us with an integrative framework (Fig. 5.1) that begins to elucidate the unique nature of engaging with citizen science data as a three-dimensional engagement. This description reveals the potential challenges of engaging with citizen science data, as learners can struggle with any of the elements detailed in the framework. It also reveals its unique potential to foster students' reasoning with each of these dimensions. In the next subsection, we describe several activities that can mediate some of these challenges and scaffold students' fruitful benefitting from the pedagogical three-dimensional potential of citizen science data.

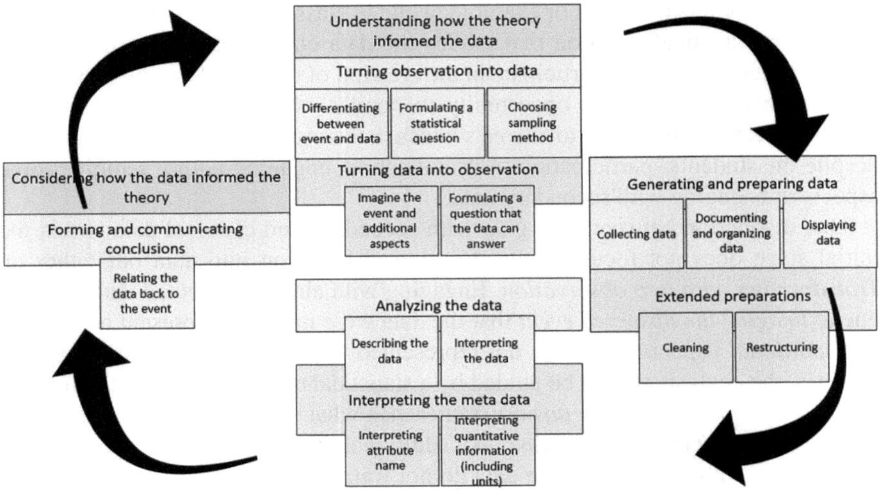

Fig. 5.1 Integrated framework of the three-dimensional engagement with citizen science data (scientific in blue; classical data in yellow; and secondary data in green)

5.3 Designing to Support Novices' Three-Dimensional Engagement with Citizen Science Data

In this section, we suggest several types of activities and approaches that can support various aspects of students' three-dimensional engagement with citizen science data. These are based on prior literature and our own research experience. Due to the complexities that characterize citizen science data, we advocate for extended educational designs that gradually scaffold the students' engagement with the data and will provide an illustrative example of such an extended educational design in the following section. However, to cater to a variety of educational goals or restrictions, we clarify the specific potential contribution of each suggestion so that one can choose to implement those that are relevant to their goals. Specifically, we structured this section according to our integrated framework and introduce our suggestions according to the elements of the framework that are most relevant to it.

5.3.1 *Activities That Can Support the Scientific Dimension of Engaging with Citizen Science Data*

5.3.1.1 Introduction Activities: Understanding How the Theory Informed the Data

In many cases in which students participate in citizen science projects, particularly when the educational objective focuses on introducing the students to the project's scientific context, it is customary to open with one or more introductory activities. These can take a variety of forms, ranging from more passive participation (e.g., listening to a lecture) to more active forms (e.g., having students read short articles in small groups with the purpose of later teaching their fellow classmates the scientific information discussed in each group's article). As in other educational contexts (e.g., Pratton & Hales, 1986), active participation has been found to be more beneficial, eliciting higher levels of engagement, agency, and motivation. We also encourage the inclusion of the project's scientists and if possible, a visit to their laboratories to provide the students with an authentic experience of the scientists' work (Tsybulsky et al., 2018a, b; Tsybulsky, 2019). Whatever the form, these activities tend to focus on the scientific knowledge that students are expected to construct and the information that is considered to be necessary for their engagement in the project (e.g., learning about the measurement tools they will use to collect data, Dvir & Ben-Zvi, 2023).

While we agree that understanding the context of their future data investigations is an important objective (Wild & Pfannkuch, 1999), we urge the consideration of three potential pitfalls. First, we caution that an emphasis on what is already "known"[1] about the phenomenon can potentially reify various common naïve views of the Nature of Scientific (NOS) that students have been reported to hold. These include the view of scientific knowledge as fixed and definite and the view of science as an authoritative system that is inaccessible to novice students (Tsybulsky, 2018). These can be mediated by emphasizing the complementary aspects of these views, such as emphasizing the tentative nature of scientific knowledge (e.g., the fact that new data can always potentially refute prior insight, Aridor et al., 2023) and explicitly discussing the potential contribution that the students, despite being novices, can have to science through their participation in the project.

Second, to support students' understanding of *how the theory informed the data,* it will likely be beneficial to be explicit about the *theory* behind the examined scientific phenomenon. Furthermore, we note that knowledge about the scientific contents that serve as the backdrop for the current research (even when discussing a current theory that explains them) is not necessarily the same as knowledge about

[1] This wording is inspired by an experience we recently had. One of the researchers leading a Citizen science project that we collaborate with unintentionally used the formulation "we know that…" numerous times when introducing her research findings to the students. At the end of this discussion, many of the students who intended to participate in the project were puzzled about the need for further study "if everything is already known."

what motivated and informed it. Consider the difference between discussing what had been previously examined and found, as opposed to discussing what questions remain, the limitations of current theories and insight, and what future findings might challenge or refute them.

Lastly, we also encourage the consideration of explicit discussions on how current theories informed the data collection protocol adopted by the project (e.g., what data are intended to be collected? How? Why?) or other activities that make the protocol and considerations that inspired it more transparent. For example, working in small groups, students can be asked to create their own data collection protocols, introduce them to their peers and discuss the affordance and limitations of each, and ultimately compare them with the protocol that the scientists of the project have chosen.

5.3.1.2 Concluding Activities: Considering How the Data Informed Theory

While students' participation in citizen science projects varies, in many cases, their main contribution is to the data collection, and they subsequently have limited engagement with the data that they helped generate (Golumbic et al., 2020). This implies that they have limited opportunities to also consider *how these data relate to the theories* that inspired the research they contributed to, thus missing the opportunity to benefit from this aspect of engaging with citizen science data. We, therefore, advocate conducting a concluding activity that encourages the students to revisit the initial theories and compare them to their data-based insight. For example, the concluding activity can take the form of a poster session in which the students create visual displays of the initial theory and their findings to introduce these to peers from other classes or their families. One of the potential benefits of such an activity is supporting the students to extend the same initial naïve NOS views we discussed in the prior subsection.

We note that even if the students do have the opportunity to meaningfully engage with the citizen science data that they help generate, they do not inherently relate their data-based insight to scientific theory. This claim is supported both by prior literature on challenges that students can have in relating *any* data back to the examined phenomenon (Konold & Higgins, 2003) and also by our own experience with students examining citizen science data. In fact, students can experience tensions relating to the differences between the disciplinary cultures of statistics and science and might need support to negotiate seemingly conflicting statistical and scientific views (Aridor et al., 2023). Concluding activities can be designed to support these negotiations, for example, asking students to develop a scientific model that explains what they statistically inferred from the data.

5.3.2 *Activities That Can Support the Classical Data Dimension of Engaging with Citizen Science Data*

There are some benefits to having students, after completing the introduction activity, immediately engage with the big data sets that are often generated by the project (e.g., appreciating the sense of being "awash in data" but developing ways to still find the shore, Erickson, 2017). However, by doing so, learners are provided with limited opportunities to reason with some of the core statistical ideas that are at the heart of any data investigation. Learners often need time and repeated experience to gradually develop their reasoning with distributions (Bakker & Gravemeijer, 2004), uncertainty (Braham & Ben-Zvi, 2019), statistical inference (Makar et al., 2011), or statistical modeling (Dvir & Ben-Zvi, 2023). Depending on the learning objectives, we advocate to initiate the students' data investigations with more limited data, for example, initially examining the data that they themselves have generated to scaffold their subsequent engagement with the project's big data. This allows the inclusion of various activities that have been designed and examined to support novices' reasoning with classical data and can enrich their engagement with the classical data dimension of engaging with citizen science data.

We provide here a few examples of such activities and design considerations:

1. *The growing samples heuristic* (Bakker & Gravemeijer, 2004): This approach advocates having learners conduct multiple classical data investigation activities of gradually larger samples. This gradual process has been known to foster more mature data descriptions and interpretations, as well as more mature views of uncertainty related to sampling and inference
2. *A data investigation cycle posters activity*: A hands-on data collection activity inspired by the data investigation cycle (e.g., Wild & Pfannkuch, 1999) that allows students to participate in all of its stages. Working in small groups, often under a given theme, each group is asked to formulate a research question and envision the data that they will need to address it. Students are told that they will collect the data from their peers and use the poster to communicate what they discover to their classmates. Thus, the students are encouraged to come up with a data collection protocol, decide how to document and then organize their data, and display it on their posters in means that they would explain to their classmates what they have learned from the data. Partial or full renditions of this activity can be incorporated depending on the educational objectives. The activity is particularly beneficial to engaging students with challenges related to formulating statistical questions (as often, their initial formulations are unclear or problematic to address). It also encourages the construction of various creative data representations, setting the ground for discussion regarding the affordances and limitations of each form of data display.
3. *Informal Statistical Inference* (ISI, Makar et al., 2011): According to this approach, the goal of the data investigation should not merely be to describe what can be seen in the data but rather to formulate data-based claims that go

beyond the data and generalize to a larger population. This approach has been known to foster students' reasoning with uncertainty related to sampling and inference.

4. *Informal Statistical Modeling* (Dvir & Ben-Zvi, 2023): According to this approach, the context of data investigations can serve as fruitful grounds to foster students' creation, evaluation and refinement of various forms of models (e.g., data models that represent what students see in the data; conjecture models that represent what the students hypothesize about the larger population, Dvir & Ben-Zvi, 2018). Modeling is not only a central statistical practice but has recently become the focus of various state-of-the-art pedagogies (Lehrer & English, 2018). Furthermore, modeling is also at the heart of many of the newer data science practices (e.g., predictive modeling), as well as the newer data-driven scientific practices. Therefore, incorporating modeling activities (e.g., designing and running a dynamic model that can generate simulated data repeatedly to allow novices to experience and evaluate sampling variation, Lehrer, 2017) can concurrently support novices' engagement with also some of the aspects of scientific and secondary dimensions of citizen science data.
5. *Incorporating technological tools*: The affordances of utilizing technological tools that facilitate the quick construction and transition between multiple forms of data displays have been discussed extensively. These include freeing the learner from laborious tasks and calculations (Cobb, 2007; Rubin & Hammerman, 2006) so that they can focus on reasoning with a wider array of data representations (Arcavi, 2003). Various tools have been specifically tailored for the use of young students (e.g., TinkerPlots, Konold & Miller, 2015) who have little or no prior experience with data.
6. *Purpose and utility* (Ainley et al., 2006): This principle calls for making explicit the overall purpose of the students' participation in the citizen science project and the utility of each task that they participate in promoting this purpose. By doing so, one can foster higher levels of internal motivation and elicit deeper engagement in each activity.

5.3.3 Activities That Can Support the Secondary Data Dimension of Engaging with Citizen Science Data

We contend that considering this dimension is particularly consequential as many challenges of engaging with citizen science data are related to and warrant intentionally design support.

5.3.3.1 Turning Data into Observation and Interpreting Metadata

Although the students often had the opportunity to participate in the data collection, the wide variety of attributes in the data might still be challenging to navigate. Furthermore, the aspects of the metadata, such as the meaning of the pre-chosen names of the attributes or how they were collected, might not be clear from the final data set that the students are expected to engage with (Rubin, 2021). This can be partly mediated by providing the students with a legend that explains the meaning of each attribute, how it was collected, the units used, and, if necessary, an indication of the type of attribute (e.g., categorical). We also suggest adding specific instructions to the first data investigation that encourage students to first examine each of the attributes separately (e.g., create a distribution for each attribute and examine what can be learned about the attribute from it). This addition has several affordances. First, it supports students' engagement with aspects related to classical data (e.g., reason with distributions, Bakker & Gravemeijer, 2004) and supports their identifying 'interesting' properties that can inspire the research questions they would like to explore (e.g., Manor & Ben-Zvi, 2017). Second, it imitates aspects of formal data science practices, often beginning with a similar examination of the pre-existing data set. In formal practice, the familiarity process also includes the identification of any problems in the data and informing the data scientist on what cleaning actions and additional preparations are needed. The latter is another reason we advocate for these additions, as the students, too, are likely to begin to identify problems in the data that need to be addressed, which can set the ground for the extended preparations activity we introduce next.

5.3.3.2 Extended Preparations

As discussed earlier, collected by multiple untrained individuals, citizen science data provides unique opportunities to engage learners with practices related to repairing and preparing data (e.g., Erickson et al., 2019). In many cases, students are presented with data that has already been cleaned, simplified, and prepared, as messy or 'dirty' data (Chu et al., 2016) are challenging to handle. Again, depending on the learning objectives, to maximize the pedagogical potential of citizen science data, we advocate for the incorporation of the students in the process of cleaning and preparing the data. Activities that facilitate this incorporation can take the form of class discussions regarding any problem that the students have identified in the data and suggestions that they have about how to address them. If suitable for the learning objectives, the students themselves can also engage in implementing some or all of their solutions on excel files of raw data or using other technological tools that offer more visual data displays. Otherwise, the teacher or researcher can implement the students' suggested cleaning actions and data moves.

We also note that even if an activity is dedicated to this process, it is reasonable that as the students continue their data investigation, they will encounter and acknowledge additional issues that can be addressed by additional data preparations. For example, when formulating a research question, students might become interested in combined attributes or discover the need to convert an attribute's type (e.g., from categorical to numerical). These can serve as additional opportunities to engage students with additional cleaning actions or data moves. We therefore suggest including ongoing prompts to encourage students to seek, articulate, and reason with any issues or problems they notice in the data. This, as well as all of the suggested supportive design aspects listed above, inspired an extended educational design we recently implemented. In the following section, we introduce this design to illustrate how these various design considerations can be incorporated into one cohesive activity sequence to maximize learners' citizen science data experience.

5.4 Illustrative Example: The Extended "Sleep One-Third of Your Life" Activity Sequence

The activity sequence we offer as an illustrative example for implementing the various suggestions detailed in the former section was tailored to accompany a tenth-grade classroom (15 students majoring in data analysis, ages 15–16) participation in the "Sleep One-Third of Your Life" citizen science project. The sequence was designed and recently implemented as part of a collaboration between a biology teaching research team and a statistics education (Connections) longitudinal design and research project, as part of the Taking Citizen Science to School (TCSS) center. The goals of the Citizen Science Sleep project are to: (1) describe the quality and quantity of Israeli teenagers' sleep; (2) gain additional insight on known factors that affect teenagers' sleeping patterns; and (3) discover new potential factors based on a large-scale data collection. The latter reflects the data-driven scientific practices that the scientists of the project employ. The project-generated data is collected by students' filling out ready-made questionnaires that constitute a "sleep journal" about their sleep habits for two weeks (14 nights). Motivated by goal three, the sleep scientists also promote the collection of any additional attributes that the students believe can be used to either describe their sleep patterns or serve as factors that might affect (or be affected by) their sleep. Thus far, the large project-generated data set consists of ~17,000 cases (nights) and includes a large variety (~60) of pre-chosen and student-suggested attributes. The ultimate goal of the nine-lesson-long learning activity we designed was to support students' engagement with this complex data set, in ways that maximize their reasoning with all three of its characteristic dimensions (scientific, classical statistics and secondary). In the following we describe each of the activities, and emphasize the design considerations that inspired it. Figure 5.2 provides an overview of both the activities and the design considerations.

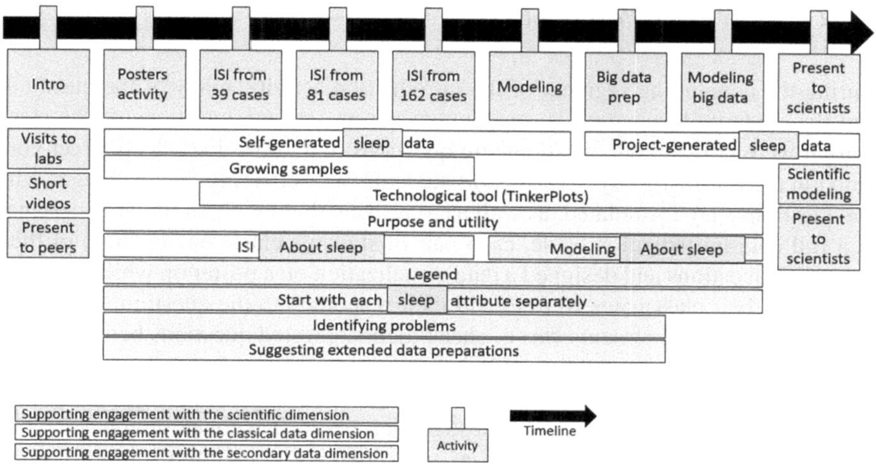

Fig. 5.2 An overview of the Sleep activity sequence and the design suggestions that were implemented in each of the activities

5.4.1 Activity 1: Introduction Activity

The introduction activity consisted of two parts. The first was a *visit* to the sleep scientists' laboratories. During the visit the students met the scientists and were provided with potent experiences of their actual scientific practices. The visit also included a brief lecture from one of the Sleep scientists and the education researchers who would accompany the students throughout their participation. The lecture introduced the students to citizen science and the Sleep project. The second part of the introduction activity was designed to introduce the students to current sleep-related theories in ways that will later make how these informed the data collection (which the students will participate in) more transparent. In this part, the classroom was divided into small groups, each provided with a different short video about known sleep theories (e.g., the effect of blue light on sleep patterns). Each group was tasked with studying the video and creating an educational slide on a shared Google presentation to *teach the rest of their classmates* about the theory they had learned. To finalize the activity, the students were introduced to the sleep journals they will begin to fill out daily, and working in pairs, the students were asked to suggest additional questions they would like to add to those pre-chosen by the sleep scientists.

5.4.2 Activity 2: Poster Activity

During the concluding segment of the introduction activity, many of the questions that the students formulated to supplement the pre-chosen questions of the sleep journal were not suitable questionnaire questions (e.g., how does sleep affect your mood in the following day?) To support students in discovering the challenges with the questions they formulated, as well as to provide them with an initial experience of a full data investigation cycle, each pair of students chose one or more of their suggested questions and designed a data visualization on a poster on which they and the rest of their classmates were to add their responses to the question/s. The *purpose* that guided the students was to check their suggested questions before including them in the sleep journal they will soon begin to fill out. After collecting the data (sample size 15), each pair was asked to analyze the data, form ISIs about Israeli teenagers' sleep habits and present these to their peers. In a concluding whole class discussion, the "problematic" questions were identified, and their initial formulations were refined. The revised questions were added by the research team to the sleep journal, and the students began filling out their journals the following morning. Additional *problems* with the data themselves were also discussed (e.g., deciphering the meaning of the name of an attribute such as "What does 'mood' mean?"), as well as additional suggestions on how to address these challenges. The scientific *sleep* context was omnipresent, for example influencing what the students considered to be "interesting" questions or results.

5.4.3 Activities 3–5: Student-Generated Data Investigations

To facilitate a gradual introduction to the project-generated data, the students then engaged in three additional small data investigations that followed a similar protocol with *gradually larger data sets*: the data that the students collected about their classroom with the sleep journal during the first three days of data collection (39 nights or cases), the first week (81 nights) and finally the two-week data (162 nights). In each activity, the students were asked to formulate a research question (now based on the already existing data) about the relation between two attributes, draw their conjectures, and organize their data to examine *each of their chosen attributes separately* using the *technological tool* TinkerPlots (Konold & Miller, 2015), and then organize their data to examine the relation between the two attributes, again using TinkerPlots. The students were then asked to describe what they saw in the data and formulate *ISIs* based on what they saw (and their prior knowledge) about the sleep habits of Israeli teenagers. They were also asked to revise their initial drawings of their conjecture according to their current ISI. After presenting the different investigations to their peers, the students were asked to choose a shared research question which they will all continue exploring with the large project-generated data set. We note that during all of the data investigations (student and project-generated) the students were provided with a *legend* that listed all of the attributes' names, how each was defined, the question from the sleep journal that elicited it, and the units it was measured with (e.g., minutes).

The *purpose* that guided the students' participation was to help the sleep scientists discover new relations in the data as soon as possible using the limited data that the classroom had collected at each stage. Students *utilized* the insight they gained in the first and second data activities as they continued to examine additional data each time, often perusing the same research question throughout the three data investigations.

Throughout the activities, students were explicitly encouraged to articulate any challenges or *problems* they encountered with the data—these included questions regarding the accuracy of the data (e.g., "Did someone really know that they fell asleep exactly in 22 minutes?"), deciphering the meaning of a pre-chosen attribute name (e.g., "What does Fall_asleep mean?"), pre-chosen units (e.g., "22 what?"), working with difficult types of attributes (e.g., the time 00:15:00, 15 min past midnight, is actually later than 23:00:00, 11 p.m., but would appear as 'smaller,' i.e., more to the left, when plotting the data) identifying missing or erroneous data (e.g., "no way it took her 420 minutes to fall asleep!"). The students were also encouraged to suggest means to overcome these challenges through *extended data preparations* (e.g., omit or correct erroneous cases; transform hourly categorical data to numerical data). We also note that during all three data activities the scientific *sleep* context was omnipresent as before, reflected in the questions and results that the students considered to be "interesting" as well as in their ISIs that often included some form of causal scientific explanation for the relation found.

5.4.4 Activity 6: Modeling

As a final preparation for transitioning to the large project-generated data set and the practice of predictive *modeling* that is more aligned with authentic big data practices, the real data investigations were followed with an initial modeling activity. The students were asked to use an additional *TinkerPlots* tool, the Sampler, that allows them, working in pairs, to design a dynamic model of their final ISI about Israeli teenagers' sleep habits. The *purpose* that was introduced for the activity was to use a new *technological tool* to express their current ISIs (after some were quite reluctant to draw them), which they will later examine and might further revise according to what they will learn from the large data set. To test their Sampler model design and refine it, if necessary, the students generated samples size 162 from the Sampler model that they created and compared them with the real 162-sized sample of the students-generated data. When the students were satisfied with their design, they were asked to generate additional simulates samples size 162 and compare between them—providing a potent experience of sampling variability. The intended design also included the creating of a sampling distribution of chosen statistics to further evaluate the sampling variability but due to time constraints this was not implemented with the students.

5.4.5 Activities 7–8: Project-Generated Data Investigations

Equipped with initial models, the students began exploring the larger project-generated data set over the span of two lessons. Each lesson was devoted to a different aspect of engaging with the data: extended data preparations (activity 7) and predictive modeling (activity 8). Activity 7 began with the students' exploring the large data set using Excel as their *technological tool.* The purpose of the activity was to identify any *problems* with the data (e.g., identifying missing or erroneous data), suggest means to overcome them through *extended data preparations* and implement these suggestions. The students were guided to focus on the two attributes of the shared research question. The *purpose* that was introduced was to "clean" the data so they would be more *useful* when the students later examine them. The students' actions included attending to erroneous cases either by omitting them or by correcting them according to additional information they had. These corrections were often sophisticated, for example, correcting an entry in the Net_sleep_time attribute indicating how many hours the teenager slept based on data about the time they fell asleep and the time that they woke up and subtracting information about total awakening duration. Midway through the activity the students were also given a TinkerPlots file of the data to facilitate creating visual displays of each of the two focal attributes (separately). The students soon discovered that the visualizations allowed easier identification of suspicious outliers or errors.

Having cleaned the data, the students began examining a partial data set (~4000 cases, due to TinkerPlots limitation as well as due to the fact that one of their focal attributes was not pre-chosen by the scientists and so only about 5000 cases included it) in activity 8. Working in pairs, the students were asked to review (and if needed also revise) their conjectured Sampler models, organize the project-generate data to examine each of the two shared focal *attributes separately* using the *technological tool* TinkerPlots (Konold & Miller, 2015), and then organize their data to examine the relation between the two attributes, again using TinkerPlots. The students were then asked to describe what they saw in the data and revise their Sampler model accordingly. As in prior data investigations, the *purpose* that guided the students' participation was again to aid the sleep scientists by discovering interesting relations. However, as in this activity all of the pairs were examining the same research question, their goal was to ultimately choose 'the best' model out of those created by all of the pairs. The activity was therefore framed as a "model competition," and the pairs were asked to develop means to convince their peer to choose their own model as the winner.

The competition was conducted in two stages. First, the pairs were grouped into two groups (three pairs in one, and four pairs in the other), and each group had to choose one of the three-four models that its pairs had developed. The main selection criteria that the students emergently suggested was what can be considered as 'model fit'—or how well does the model exactly fits the data. In the second stage, each of the two groups introduced their chosen model and had to convince the other group to adopt it. The criteria of prediction were then introduced, and the students were given an additional data set (~1000) to test which of the two rivaling models more accurately predicted the new data. This innovative activity design was inspired by the practice of using an algorithm for finding a model that offers predetermined descriptive and predictive levels that characterize both data science secondary data analysis as well as scientific data-driven practices. We note again that the scientific *sleep* context was omnipresent throughout these activities.

5.4.6 Concluding Activity

The concluding activity was conducted in a celebratory manner, in which the students presented their work to the Sleep scientists as well as a cohort of esteemed school administrators, representatives from the education municipality, and additional TCSS researchers. The original design intended to ask the students to create *scientific models*, representing scientific theories that explain the winning predictive (statistical) model. Due to time constraints this was ultimately conducted with the sleep scientists during the students' presentations. The students were proud to learn that they actually have discovered a relation in their data that was thus far only accounted for with adults, and that there is not yet an accepted scientific theory that fully explains it. The sleep scientists assured the students that they will now further examine the data and suggest additional research to better understand the phenomenon that the students discovered. This serves as an example of how students' meaningful engagement with citizen science data can be highly beneficial to students' learning and empowerment, as well as science.

5.5 Conclusions: Citizen Science Data as a Gateway to the Newer Data-Driven Scientific Practices

In this chapter, we have introduced our integrated framework that characterizes the engagement with citizen science data as three-dimensional and illustrated how it can inform educational designs on how to benefit from its unique pedagogical potential. This foundational feature of citizen science has various implications. First, meaningfully engaging learners with the data that they help generate in ways that scaffold their engagement and support their overcoming the inherent challenges can provide a potent experience of authentic practices related to classical statistics, data science, and science. Although discussed here in the specific context of citizen science, the integration of these practices also characterizes the new data-driven practices that seek to generate novel scientific insight from already existing data sets. As part of these practices, these data are subject to secondary data analysis, which builds on classical data practices while extending them into broader interpretative frameworks. Importantly, the epistemological status of these data differs depending on one's perspective: from a positivist standpoint, data provide an objective record from which patterns emerge; from a critical realist perspective, data are not neutral artifacts but rather shaped by the methods, contexts, and theoretical lenses through which they are collected and analyzed. The bidirectional relation between data and scientific theory also characterizes these newer data practices, as existing theories offer predictions regarding the data expected to be generated if the theory were true (even if they were collected serendipitously). These data, in turn, do not simply confirm or refute a theory but interact with it, informing, extending, or refining scientific understanding. One central difference, however, is that while other authentic, ongoing scientific research endeavors that implement data-driven scientific practices are often inaccessible to the public and to students in particular, citizen science projects are accessible. This open engagement with data and its interpretation creates a unique pedagogical opportunity, allowing students to experience not only the technical challenges of working with real-world data but also the conceptual challenges of understanding what data mean in science. This is but one of the unique pedagogical opportunities afforded by incorporating students in citizen science, but one that is at the very frontiers of the sciences of tomorrow.

Acknowledgments This research was conducted as part of the TCSS Center of Excellence, which was supported by the Israel Science Foundation (grant no. 2678/17). In addition, the research was supported by the Israel Science Foundation (grant no. 1345/22).

References

Ainley, M., Pratt, D., & Hansen, A. (2006). Connecting engagement and focus in pedagogic task design. *British Educational Research Journal, 32*(1), 23–38. https://doi.org/10.1080/01411920500401971

Arcavi, A. (2003). The role of visual representations in the learning of mathematics. *Educational Studies in Mathematics, 52*(3), 215–241. https://doi.org/10.1023/A:1024312321077

Aridor, K., Dvir, M., Tsybulsky, D., & Ben-Zvi, D. (2023). Living the DReaM: The interrelations between statistical, scientific and nature of science uncertainty articulations through citizen science. *Instructional Science, 51*, 729–762. https://doi.org/10.1007/s11251-023-09626-8

Bakker, A., & Gravemeijer, K. P. (2004). Learning to reason about distribution. In D. Ben-Zvi & J. Garfield (Eds.), *The challenge of developing statistical literacy, reasoning and thinking* (pp. 147–168). Springer.

Bargagliotti, A., Franklin, C., Arnold, P., Gould, R., Johnson, S., Perez, L., & Spangler, D. (2020). *Pre-K-12 guidelines for assessment and instruction in statistics education (GAISE) report II*. American Statistical Association and National Council of Teachers of Mathematics.

Bauer, H. H. (1994). *Scientific literacy and the myth of the scientific method*. University of Illinois Press.

Ben-Zvi, D. (2004). Reasoning about data analysis. In D. Ben-Zvi & J. Garfield (Eds.), *The challenge of developing statistical literacy, reasoning and thinking* (pp. 121–145). Kluwer.

Biehler, R., Ben-Zvi, D., Bakker, A., & Makar, K. (2018). Technology for enhancing statistical reasoning at the school level. In K. Makar, B. de Sousa, & R. Gould (Eds.), *International handbook of research in statistics education* (pp. 139–192). Springer.

Braham, H. M., & Ben-Zvi, D. (2019). Design for reasoning with uncertainty. In G. Burrill & D. Ben-Zvi (Eds.), *Topics and trends in current statistics education research: International perspectives* (pp. 97–121). https://doi.org/10.1007/978-3-030-03472-6_5

Cao, L. (2018). Data science: A comprehensive overview. *ACM Computing Surveys, 50*(3), 1–42. https://doi.org/10.1145/3076253

Chalmers, A. F. (2013). *What is this thing called science?* (4th ed.). Hackett Publishing Company.

Chu, X., Ilyas, I. F., Krishnan, S., & Wang, J. (2016). Data cleaning: Overview and emerging challenges. In *Proceedings of the 2016 international conference on management of data* (pp. 2201–2206).

Cobb, G. W. (2007). The introductory statistics course: A Ptolemaic curriculum? *Technology Innovations in Statistics Education, 1*(1). https://doi.org/10.5070/T511000028

Dvir, M., & Ben-Zvi, D. (2018). The role of model comparison in young learners' reasoning with statistical models and modeling. *ZDM, 50*, 1183–1196.

Dvir, M., & Ben-Zvi, D. (2023). Informal statistical models and modeling. *Mathematical Thinking and Learning, 25*(1), 79–99. https://doi.org/10.1080/10986065.2021.1925842

Erickson, T. (2017, Feb 21). Smelling like data science. https://bestcase.wordpress.com/2017/02/21/smelling-like-data-science/

Erickson, T., Wilkerson, M., Finzer, W., & Reichsman, F. (2019). Data moves. *Technology Innovations in Statistics Education, 12*(1). https://doi.org/10.5070/T5121038001

Friel, S. N., & Bright, G. W. (1998). Teach-stat: A model for professional development in data analysis and statistics for teachers K-12. In S. P. Lajoie (Ed.), *Reflections on statistics: Learning, teaching, and assessment in grades K-12* (pp. 89–117). Lawrence Erlbaum Associates.

Gasparatou, R. (2017). Scientism and scientific thinking. *Science & Education, 26*(7–9), 799–812. https://doi.org/10.1007/s11191-017-9931-1

Golumbic, Y. N., Baram-Tsabari, A., & Koichu, B. (2020). Engagement and communication features of scientifically successful citizen science projects. *Environmental Communication, 14*(4), 465–480. https://doi.org/10.1080/17524032.2019.1687101

Gould, R. (2021). Toward data-scientific thinking. *Teaching Statistics, 43*, S11–S22. https://doi.org/10.1111/test.12267

Kanari, Z., & Millar, R. (2004). Reasoning from data: How students collect and interpret data in science investigations. *Journal of Research in Science Teaching, 41*(7), 748–769. https://doi.org/10.1002/tea.20020

Kelling, S., Fink, D., La Sorte, F. A., Johnston, A., Bruns, N. E., & Hochachka, W. M. (2015). Taking a "big data" approach to data quality in a citizen science project. *Ambio, 44*(S4), 601–611. https://doi.org/10.1007/s13280-015-0710-4

Kjelvik, M. K., & Schultheis, E. H. (2019). Getting messy with authentic data: Exploring the potential of using data from scientific research to support student data literacy. *CBE—Life Sciences Education, 18*(2), es2. https://doi.org/10.1187/cbe.18-02-0023

Konold, C., & Higgins, T. (2003). Reasoning about data. *A research companion to principles and standards for school mathematics, 193215*.

Konold, C., & Miller, C. (2015). *TinkerPlots* (Version 2.3.1) [Computer software]. University at Massachusetts. Online: http://www.tinkerplots.com/

Lehrer, R. (2017). Modeling signal-noise processes supports student construction of a hierarchical image of sample. *Statistics Education Research Journal, 16*(2), 64–85.

Lehrer, R., & English, L. (2018). Introducing children to modeling variability. In *International handbook of research in statistics education* (pp. 229–260). Springer International Publishing.

Makar, K., Bakker, A., & Ben-Zvi, D. (2011). The reasoning behind informal statistical inference. *Mathematical Thinking and Learning, 13*(1–2), 152–173. https://doi.org/10.1080/10986065.2011.538301

Manor, H., & Ben-Zvi, D. (2017). Students' emergent articulations of statistical models and modeling in making informal statistical inferences. *Statistics Education Research Journal, 16*(2), 116–143. https://doi.org/10.52041/serj.v16i2.187

McClain, K., Cobb, P., & Gravemeijer, K. P. E. (2000). Supporting students' ways of reasoning about data. In M. J. Burke & F. R. Curcio (Eds.), *Learning mathematics for a new century* (2001 yearbook) (pp. 174–187). National Council of Teachers of Mathematics.

Popper, K. R. (1963). *Conjectures and refutations: The growth of scientific knowledge*. Routledge.

Pratton, J., & Hales, L. W. (1986). The effects of active participation on student learning. *Journal of Educational Research, 79*(4), 210–215. https://doi.org/10.1080/00220671.1986.10885679

Rubin, A. (2021). What to consider when we consider data. *Teaching Statistics, 43*, S23–S33. https://doi.org/10.1111/test.12275

Rubin, A., & Hammerman, J. K. (2006). Understanding data through new software representations. In G. Burrill & P. C. Elliott (Eds.), *Thinking and reasoning with data and chance: 68th NCTM yearbook* (pp. 241–256). National Council of Teachers of Mathematics.

Schuttler, S. G., Sorensen, A. E., Jordan, R. C., Cooper, C., & Shwartz, A. (2019). Bridging the nature gap: Can citizen science engage the unengaged in nature and science? *Science Advances, 5*(7), eaav0767. https://doi.org/10.1002/fee.1826

Tsybulsky, D. (2018). Comparing the impact of two science-as-inquiry methods on the NOS understanding of high-school biology students. *Science & Education, 27*(7), 661–683. https://doi.org/10.1007/s11191-018-0001-0

Tsybulsky, D. (2019). Students meet authentic science: The valence and foci of experiences reported by high-school biology students regarding their participation in a science outreach programme. *International Journal of Science Education, 41*(5), 567–585. https://doi.org/10.1080/09500693.2019.1570380

Tsybulsky, D., Dodick, J., & Camhi, J. (2018a). High-school students in university research labs? Implementing an outreach model based on the 'science as inquiry' approach. *Journal of Biological Education, 52*(4), 415–428. https://doi.org/10.1080/00219266.2017.1403360

Tsybulsky, D., Dodick, J., & Camhi, J. (2018b). The effect of field trips to university research labs on Israeli high school students' NOS understanding. *Research in Science Education, 48*(6), 1247–1272. https://doi.org/10.1007/s11165-016-9601-3

Wild, C. J., & Pfannkuch, M. (1999). Statistical thinking in empirical enquiry. *International Statistical Review, 67*(3), 223–248. https://doi.org/10.1111/j.1751-5823.1999.tb00442.x

Chapter 6
Teaching Ecology and Inquiry in the Field Through Citizen Science

Nirit Lavie Alon, Tali Tal, Hila Shefet Barkae, and Lavi Oren

6.1 Introduction

This chapter presents the stories of two exemplary biology teachers who teach at the middle school level (grades 8 and 9) in Israel. Each developed a science curriculum that included out-of-school citizen science projects. Both teachers were eager to teach in a new way and were committed to providing the best education to their students while also engaging in personal and professional development. Both teachers developed and tested a unique citizen science-based curriculum as part of their graduate work. The first two authors are the PIs and advisors of the two teachers, and the second and third authors are the teachers themselves.

Both teachers are in their 30s. Lavi, a male teacher, taught 8th grade science to a class of students with low academic achievements in the geographical periphery which is characterized by the lower socio-economic status of its residents. The school serves many immigrant students from countries such as Russia, Ukraine, Ethiopia, and the United States. The combination of recent immigrant populations and low socio-economic status results in a student body that have not fully mastered Hebrew and whose academic achievements are low, especially in science.

N. L. Alon (✉) · H. S. Barkae · L. Oren
Faculty of Education in Science and Technology, Technion Israel Institute of Technology, Haifa, Israel
e-mail: nirita@ed.technion.ac.il

T. Tal
Faculty of Education in Science and Technology, Technion Israel Institute of Technology, Samuel Neaman Institute, Haifa, Israel
e-mail: rtal@ed.technion.ac.il

D. Tsybulsky, T. Tal (eds.), *Citizen Science Contributions to Biology and Environmental Education*, Contributions from Biology Education Research,
https://doi.org/10.1007/978-3-031-99298-8_6

Hila, a female teacher, taught the honors 9th grade science class at a school in the center of the country. Her students come from middle to upper-middle class backgrounds. The school has received several prestigious educational awards.

In some Israeli schools, tracks are determined in 7th grade, based on achievements. Certain subjects are split into high-achieving and low-achieving classes. Lavi taught the lowest-achieving class in science, whereas Hila taught the highest-achieving students.

Teachers in Israel are required to develop lesson plans that adhere to the national curriculum that covers detailed content specifications, learning goals, expected learning outcomes, and the associated skills and practices. They have some autonomy in terms of content and teaching methods, provided they are aligned with the national curriculum. Both Hila and Lavi chose to teach the topic of ecological systems (biological diversity in nature, interrelationships between organisms and their environment, dynamics of ecological systems, and human involvement in the environment) through out-of-school learning that integrated citizen science.

Lavi, who taught the 8th grade class, aimed to enrich the learning environment for his students, allow them to experience science in more informal settings, familiarize themselves with their surroundings, and cultivate their interest in science. He developed a curriculum that primarily focused on studying ecology outdoors, which was concretized by a citizen science project that involved monitoring rare plant species in the area. With the help of his advisors, he worked in collaboration with the Nature and Parks Authority to guide his students to survey these rare species in a nearby grove.

Hila's objective was to avoid prescriptive inquiry learning by taking part in an authentic research experience that would be relevant to her students and make a meaningful contribution to science. She contacted an ecologist from the Society for the Protection of Nature who pointed to a "mystery" related to fertilization of a rare species of Iris (*Iris atropurpurea*). Hila, then, designed a curriculum centered around a student-led research project investigating the Iris population.

They both then conducted empirical assessments of the impact of their programs on their students.[1] Lavi examined the impact of the citizen science project on academic achievement and attitudes toward science. Hila examined the characteristics of a "bottom up" citizen science project and the students' learning outcomes. Both studies employed a mixed method approach that combined qualitative instruments such as observations, document analysis and interviews with quantitative pre- and post-project questionnaires. This chapter presents insights from these two examples of tailored instructional programs that combine citizen science with science education and illustrates how citizen science design principles can be implemented. we point to the contribution of these two citizen science initiatives to the students.

[1]Both studies received IRB approval (2021-033) and ethical approval from The Ministry of Education # 11781; # 10440

6.2 Background

In recent years, growing numbers of projects and participants have engaged in citizen science, in particular in educational settings (Edwards, 2014). Models of citizen science differ in the extent of public involvement and influence, and the way the public participates (Palumbo et al., 2022; Pocock et al., 2017).

There are a variety of typologies for classifying and understanding citizen science participation. One prominent classification of public involvement was proposed by Haklay (2013):

- Crowdsourcing: the most basic level of participation that involves the provision of resources with minimal cognitive engagement.
- Distributed intelligence: Taps cognitive abilities. Participants are given brief training and then collect data or carry out a simple interpretation activity.
- Community science (Participatory science): Involves the participants in defining the problem and devising the data collection method. The participants then engage in data collection but experts analyze and interpret the results.
- Collaborative science (Extreme citizen science): the most highly integrated level of participation where participants decide which scientific problems to work on, the nature of the data collection, and the analysis and publication or utilization of results.

The two examples presented here are instances of Collaborative (sometimes termed Co-created; see Bonney et al., 2009) Citizen Science in which teachers initiate the study and develop it in conjunction with experts and scientists. In the 8th grade study, the students were only involved in the data collection (Oren et al., 2022). In the 9th grade study, the students were also involved in formulating the research questions, collecting and analyzing the data, and presenting them to the community (Tal et al., 2023).

Citizen science in schools is also on the rise. Participation in citizen science can promote knowledge, skills with values, a better understanding of the nature of science (NOS), and of how to use scientific practices. Citizen science kindles students' interest and curiosity because of its authenticity and relevance. These projects contribute to developing agency and activism, and the safeguarding of the environment. Citizen science also provides an opportunity for both students and teachers to gain self-efficacy (Bonney et al., 2009; Phillips et al., 2018; The National Academies of Sciences Engineering and Medicine, 2018).

6.2.1 Principles for the Design of Citizen Science in Schools

"Taking Citizen Science to Schools" (TCSS) is an inter-university center that brings together researchers from the University of Haifa and the Technion, both of which are located in the city of Haifa in Israel. The Center, which was started in 2017, is one of five research Centers of Excellence whose aim is to promote meaningful learning and is funded by the Israel Science Foundation and the Israel Ministry of Education. The TCSS is composed of a network of partnerships for design-based research that aims at promoting the integration of citizen science in science classes (Hod et al., 2018). The center supports the integration of citizen science projects in schools through a variety of research projects.

One of the center's research projects was to develop design principles for citizen science in schools, through the combined efforts of members of the TCSS community who do citizen science in schools including education researchers, organizations involved with citizen science and scientists who are researching various topics through citizen science, teachers who lead citizen science in their schools, students, and others interested in the field.

Design principles are general guidelines to help in the planning of teaching and the development of activities that support in-depth learning for students in different contexts, especially in complex and challenging situations (heterogeneous classrooms, outdoor learning, complex content, etc.). The principles are based on the Design Principles Database (DPD) developed in the early 2000s (Kali, 2006). The original Design Principles Database was developed by Kali (2006) with 50 design researchers in a collaborative process, to coalesce and synthesize emerging design knowledge about the use of technologies for education. Inspired by the DPD, the TCSS researchers developed a revised set of design principles for citizen science specifically in schools. This effort involved meetings with the community, workshops with a demo repository of the principles, and work with cards of the principles. It led to the creation of a new, dynamic, online collaborative platform dubbed "Insights", which is unique in that it was developed by a heterogeneous community of citizen science practitioners through storytelling (Aridor et al., 2022). In total, 27 design principles were formulated. Some are broad and suitable for any form of science learning or inquiry learning, but others are more specific to citizen science.

Despite the diversity of topics in citizen science, most projects are related to environmental research (Pocock et al., 2017), are carried out in the field, and involve outdoor learning. Thus, one of the design principles covers ways to bridge between in-class and outdoor learning.

6.2.2 *Contributions of Outdoor Learning*

Outdoor learning can take place in a variety of settings outside the classroom, such as the schoolyard and its surroundings, gardens and parks, nature parks and the urban environment, museums and visitor centers. Many studies have underscored the value of learning outcomes in outdoor environments, which include the honing of cognitive skills such as knowledge acquisition, better long-term memory, and boosted academic achievement, the development of attitudes and values, emotional skills such as regulation and self-management interpersonal/social skills, leadership, and teamwork, as well as physical and behavioral benefits (DeWitt & Storksdieck, 2008; Malone, 2008; Rennie, 2014; Tal, 2012).

Outdoor learning is at its most productive when it is based on design principles that link it to a learning sequence that includes preparation, summation, and activities that reduce unfamiliarity (Orion, 1993), uses the environment in which it takes place (Lavie Alon & Tal, 2017), includes active learning that combines head-heart-hand learning that encourages thinking and higher order thinking skills such as comparison, argumentation, and drawing conclusions (head), learning from motivation, interest, and desire and developing caring for the individual and the environment (heart), and learning in which students do, experiment, and explore on their own (hand) (Sipos et al., 2008). Other studies have shown that outdoor learning encourages social interactions between learners and between them and the facilitators. Outdoor learning allows for freedom of choice, combines storytelling and play, is relevant to everyday life, and is based on local action and contributions to the community (Ardoin et al., 2020; Tal et al., 2014).

Studies on outdoor learning in the context of citizen science have found that combining the two contributes to environmental literacy (Ballard et al., 2017; Lugg, 2007; Tal et al., 2023).

6.3 Insights and Findings

It is worth noting that both projects were held with one class, and as such, the results can be treated as preliminary. Yet, the nature of specifically tailored citizen science projects necessarily limits the number of classes. Lavi has launched his project as a result of the school policy that limits access of the low-achieving classes to the resources of the science lab or the computer lab. On the other hand, Hila was eager to do a project that extends the curriculum to accommodate her teaching to her excellent students.

6.3.1 Contribution of the Programs to the Students

In the two projects presented in this chapter, participation contributed to students' scientific knowledge of ecological topics. There was also improvement in attitudes toward science and science learning in both the lower-achieving and outstanding students. The 9th grade students who experienced a complete inquiry had hands-on opportunities to apply many scientific practices during the project.

6.3.2 Contribution of the Program to 8th Grade Low-Achieving Students

As shown in Table 6.1, the knowledge gain in ecology was significant in the low achieving and the high achieving comparison class that did not participate in the citizen science program. On the other hand, a similar comparison class of low achievers (without intervention) showed no knowledge gain.

Student attitudes toward science did not improve and in fact declined, although the drop in the intervention class was not significant. There was almost no difference between the students who studied ecology integrated with citizen science and the two comparison classes in terms of enjoyment of learning science as measured by a specific question on the pre/post questionnaire (Table 6.1). Despite these results, when asked in the interviews about the project and whether they would recommend such projects to other students, 56% of the students said the project was fun and interesting; 19% referred to social interactions such as learning to know the students better, and said it increased social cohesion; 11% referred to newly acquired knowledge and understanding. Only 9% indicated a change in attitudes towards nature and the environment, and only 6% addressed motivation and interest in learning. All the students stated they would like to take part in a project again and would recommend it to others and explained that it constituted a special environment that involved tangible, active learning that differed from conventional learning in the classroom, and allowed the students freedom of action. One student said: "In class, you feel like you have to sit and study, and if you're doing something out of the ordinary, even just a little bit, then you are immediately yelled at, you can't really say what you want to say… In the grove you can listen to music, talk to friends, but even while you're talking with friends, you are still learning".

6.3.3 Contribution of the Program to the 9th Grade Honors Students

There were 297 utterances referring to the contribution of the program in the responses of the 22 students, of which 53% expressed interest and positive attitudes towards science and the environment, 21% referred to the acquisition of research skills, 17% referred to content knowledge and to the nature of science, and 8% addressed improving their environmental awareness and behavior.

Table 6.1 Mean scores for the research class and the two comparison classes before and after studying ecology

	Mean knowledge gain (max = 32)	Mean attitude towards science gain (max = 5)	"I enjoy learning science"—Change in mean score before and after
Intervention class low achievers	7.5***	−0.8	−0.0069
Comparison class high achievers	7.8***	−2.81*	−0.4523
Comparison class low achievers	−4.1	−4.91*	−0.7167*

*p < 0.05 ***p < 0.0001

Contribution of the program to interest in science—Given that this was an honors class prior to the program, it came as no surprise that 80% of the students reported interest in science, but this rose to 95% after the program. Overall, 92% of the students stated that they would be willing to participate in this type of program again or recommend it to friends. When asked why, the majority said it was a wonderful experience and helped escape from routine (53% of the statements), while others mentioned their interest in science and the environment (30%), learning new things (10%), and acquisition of research skills (7%).

Both studies thus provided insights into how to successfully plan a local citizen science project that includes the outdoor learning of science. Below, we present the lessons learned from Hila's and Lavi's experiences in designing and implementing their programs. Both studies were conducted while the design principles were being developed. Thus their input was invaluable in crafting the final principles (Table 6.2). These design principles are based on the extensive experience of teachers, scientists, citizen science leaders and education researchers. Presently, we can link the insights from the findings of Lavi's and Hila's research to the design principles as detailed below.

The seven key principles (or prerequisites) for planning and implementing citizen science in curricula are detailed below.

(i) **Educational leadership, motivation, and initiative:** Citizen science projects require collaboration and partnerships with stakeholders such as scientists and community members, as well as coordination with the school (in terms of scheduling the activities, contacts with other teachers, coordination of activities outside the classroom, etc.), alongside the use of technological devices, new pedagogical practices, and other skills that call for educational leadership. Educational leadership is essential, and the leading teacher's personality has an impact on the way it is implemented, as well as on the recruitment of support from the principal, the teaching staff, and parents.

The two teachers presented in this chapter wanted to teach their students in a different, interesting, and productive way. They were both willing to initiate a citizen science project from its inception, write the curriculum, and experiment with innovative teaching. This corresponds to design principle "Cultivating leadership in learners through personal example" (#9).

(ii) **Knowledge of the local environment and choice of a topic for the project:** In order to design a citizen science project that affords authentic learning that contributes not only to the scientific but also to the local community, a suitable topic, question or problem near the school needs to be defined. To do so, discussions with experts or community representatives are imperative to learn about the area and identify specific issues or problems or questions which are of interest to the locals.

Lavi knew of a grove about a kilometer walk from the school. He reconnoitered the grove to see whether it would be suitable for a project. He verified whether the grove was diverse enough for learning about ecological topics; for example, whether students could study the relationships between species. He

Table 6.2 Design principles for citizen science in schools and their implementation in the two programs

Design principle	Explanation	Implementation in the 8th-grade program	Implementation in the 9th-grade program
1. Bridging between in-class and outdoor learning	Creating opportunities to connect in-class and out-of-school learning to enrich learning and its relevance. When outdoor learning is planned correctly, it encourages meaningful learning	The program included learning in the classroom and in a grove, which were connected	The program included learning in the classroom and in the field, which were connected. It also included a visit to a nature reserve and a museum in addition to the field study
2. Bridging between students, teachers, and scientists	Connecting the partners involved in the study, such as teachers and scientists in the planning of activities, and scientists and students, helps guarantee that student activities will actually make a contribution to science, while at the same time be related to the curriculum and allow students to develop scientific skills	Scientists helped the teachers develop a reporting protocol tailored to the students and interacted with the students in the field	
3. Teaching sequences that are interesting, varied, and clear to the learner	A learning process that includes a teaching sequence of a number of lessons around a central idea and/or is focused on teaching that develops the learners' interest and engagement	The programs included a series of lessons which were clearly connected. The lessons dovetailed and provided a range of activities	
4. Using visual aids to illustrate complex phenomena and ideas	Dynamic or static visual aids (such as 3D models that show the phenomenon from different angles or virtual reality technology) can clarify complex phenomena, encourage students to explore the internal and external properties of phenomena, and improve their spatial abilities	Students used a photographed catalogue of rare plant species in the grove	Students visited a natural history museum during the program and saw exhibits on nature conservation and the importance of biodiversity

(continued)

Table 6.2 (continued)

Design principle	Explanation	Implementation in the 8th-grade program	Implementation in the 9th-grade program
5. Encouraging reflection	Incorporating reflection at different points in the learning process helps students reflect on their own progress, makes their thinking visible, and generates insights about the content being learned and the learning process	The overall activity was evaluated on a question to whether they would participate in this type of a program again and recommend it to their friends	The students filled in several reflective questionnaires after the activities, and a summary questionnaire about the entire program including a question about recommending the program to others
6. Bridging between in-school learning and the community	Connecting classroom learning to the local community where the global community provides students with meaningful opportunities to participate in the advancement of society. This partnership can develop social engagement and a sense of meaning while reinforcing twenty-first century skills	The activities in the grove were shared with other teachers and parents. The school administration proudly publicized the program widely	Students met with local residents who were working to protect the irises and made joint efforts to have the area declared a nature reserve
7. Bridging between the physical space and the virtual space	Virtual space has become an integral part of people's lives, for example the use of navigation software, online payment methods, and social media. Citizen science allows students to engage in activities that bridge between physical space and the virtual space even when they are at school	Students used a dedicated app to identify plants and report on rare species	Students used shared documents to share data, create presentations, and submit reflections

(continued)

Table 6.2 (continued)

Design principle	Explanation	Implementation in the 8th-grade program	Implementation in the 9th-grade program
8. Creating opportunities for student choice	Independence in learning in general, and freedom of choice in particular, encourage internal motivation that leads to persistence, higher performance, and greater satisfaction. Opening up possibilities to choose can contribute to academic achievement and can lead to greater responsibility for personal and group learning	Students chose work teams for tasks and could wander in the grove and select the sites where they wanted to do their assignments	Students could make a variety of choices: deciding on the group's research question, deciding on team members for collaborative work, and ways and means of presenting their results and insights
9. Cultivating leadership in learners through personal examples	To cultivate leadership in students, the teaching staff needs to model leadership. This leadership can empower students, prepare them for challenges in their lives and equip them with the tools to deal with tomorrow's complex reality. These tools promote personal achievements, enable students to develop their efficacy, and examine the values governing their behavior	The teachers initiated and constructed a unique curriculum, reached out to experts, coordinated the meetings and modeled their expertise and leadership to the students	
10. Support in the representation and organization of learners' knowledge	Learning processes, especially in today's information age, may create cognitive load and/or emotional flooding. Organizing the information visually (tables, concepts map, flow charts, etc.) can help learners deal with this complexity, and delve into the content, while making connections between concepts and ideas	Students learned about food webs through visual representations and flow charts	The research process was presented as a circular flow chart where each stage was presented in depth as well as the connections between stages

(continued)

Table 6.2 (continued)

Design principle	Explanation	Implementation in the 8th-grade program	Implementation in the 9th-grade program
11. Support for the visualization of learner data	Creating tools that allow students to present data they collect in an interactive and dynamic way (for example in a table, graph, or map). Visual representations can support the understanding of the phenomena and increase students' motivation to engage in research	The students compiled the observations they collected about the rare species in a chart	The students learned to organize the data they collected in a table and transform it into graphs
12. Exposing learners to the breadth of the scientific method	Students need a grasp of the breadth of scientific methods; for example, through exposure to scientific research in different disciplines or with different methodologies. Students need to discover and have hands-on experiences of ways to advance knowledge in science, which requires responsibility in data collection, evaluation of the reliability of reports, as well as collaborative work. Open discourse and discussions by the students of the work scientists do and the NOS contribute to understanding the complexity of scientific research	The students met a scientist in the field and learned about her work, but they were not exposed to the breadth of the scientific method	The students learned about studies on irises, met a researcher in the field and learned about her work and methods. They collected data in a collective form for a collaborative database. There was an open discussion about the scientist's work and NOS and the students were asked about the importance of science and their motivations in the field. They also experienced failure and had to modify their research in light of the changing situation in the field

(continued)

Table 6.2 (continued)

Design principle	Explanation	Implementation in the 8th-grade program	Implementation in the 9th-grade program
13. Dealing with complex content	Encouraging students to engage in the study of authentic and/or relevant problems allows for a better understanding of complex and controversial issues. Such issues call for an examination of various aspects of the problem, the development of systemic thinking, encouraging students to formulate a stance and eventually to take action to address the problem	The students dealt with the issue of conservation versus development of the area surrounding the grove, and suggested ideas to conserve the area	The students discovered that some of the irises were in danger and initiated activities to prevent this: a petition and protest, signage in the area and appealing to the municipality. When they realized that urban development was inevitable, they worked to move the irises to a nearby area and declare it a reserve
14. Support for the development of scientific practices and higher order thinking	Asking questions, hypothesizing, collecting and analyzing data, engaging in problem solving, and presenting opinions through evidence-based arguments are the essence of science. Supporting these practices allows students to develop higher-order thinking and encourage the creativity and mental open-mindedness involved in scientific thinking	The students did not carry out structured research but rather made observations of phenomena in the grove, asked questions about them, and made hypotheses	The students learned scientific practices in a structured way in the classroom in dedicated lessons. They experienced asking questions, raising hypotheses, and collecting data in the field. They were given instructions and support in processing the data and drawing conclusions
15. Creation and reuse of student products	Creating artifacts (e.g. presentations, models, games) and using them as resources for further learning gives meaning to the artifacts themselves and to the process of their creation	The students did not create products that were used as a resource for further learning	The students created posters and stickers that were uploaded to a Facebook page as a protest the planned industrialization of the habitat, as well as a presentation about the learning process delivered at a parents' evening

(continued)

Table 6.2 (continued)

Design principle	Explanation	Implementation in the 8th-grade program	Implementation in the 9th-grade program
16. Combining interaction methods and media types	Designing activities that combine diverse ways of learning while utilizing technology without losing the benefits of a face-to-face learning experience	The students used the app to identify plant species and report the rare species	Students collected data in Excel, used collaborative documents and spreadsheets and wrote their reflections in online forms
17. Support for a variety of social structures	Learning in a variety of social structures (independent work, pairs, teams of different sizes and compositions, and whole class)	The activity in the field included work in groups alongside whole classwork	The study was carried out by teams of 2–3 students and there were also group tasks and whole class discussions. They also worked with community members

Note: Ten other design principles are not listed here because were less relevant to these programs

also controlled for technical issues such as how long it would take to walk to the grove, the best entrance, and places to sit. He also became aware of the grove's ecological importance as an area that contained rare plant species.

Hila met with an ecologist from an environmental NGO (The Society for the Protection of Nature in Israel) to select an area near the school and to look for organisms, phenomena, or topics that would appropriate for citizen science research (i.e., had scientific value) and suitable for investigation by students. She located reports on the flowering of *Iris* atropurpurea (a rare, protected species, that is endemic to the region, and endangered), near the school and went out to look for examples. She located several different clusters of irises and noticed that in one, fruit setting has failed, and then decided that this phenomenon could be the students' research topic: the students would collect data on irises, to explore why some irises do not fertilize, while conducting scientific research in the field and in the classroom at all stages, from raising questions to drawing conclusions.

This corresponded to two design principles: "Bridging between in-class and outdoor learning" (1) and "Bridging between in-school learning and the community" (6).

(iii) **Learning about the topic:** To lead a research or citizen science project, extensive knowledge about the topic is not necessary, but basic, preliminary information is essential. This knowledge of the topic serves to develop the curriculum and choose the right tools to collect data and conduct the study. Knowledge boosts teachers' confidence in teaching and investigating the topic.

Lavi started by searching for information about the grove on the internet. He discovered that the grove has a unique concentration of rare plant species due to its location in a habitat characterized by dense soil that is underrepresented in nature reserves and open spaces in Israel. Over the years, agricultural areas have encroached on this area, and it is threatened by urban development of a new road, infrastructure and the expansion of residential neighborhoods. The grove has been named: "The Grove of Rare Species."

Hila learned about irises from books and articles on the subject, met with iris researchers, and went on a learning tour in a shelter garden for irises to broaden her expertise. Using the internet to learn more about the topic is an example of the design principle "Bridging between the physical space and the virtual space" (7).

(iv) **Establishing contact with professionals to plan citizen science:** Citizen science involves a collaboration between scientists and community members during the different phases of the research. In the case of bottom-up citizen science, as in the two examples here, it is critical to contact professionals to gather information about the topic and plan the collection and analysis of data in a way that can be of potential use to scientists who can exploit this information, while at the same time remaining appropriate for students.

Lavi read about the rare species found in the unique habitat of the grove and as a result, he contacted experts from the Nature and Parks Authority and initiated the idea of monitoring these species with his students during their learning in the grove. He explored the grove with the experts, learned to identify the plants and their characteristics, and was given access to an app the students could use to report on the rare species there.

Hila benefited from guidance from iris researchers to develop a data collection protocol for her students. This involved how often they would go out to the field, what data they would collect, what they would measure and how. She developed a provisional plan for the study with the students and the citizen science project with their help. Both Lavi and Hila collected additional information on monitoring species in the field and applied for a permit from the Nature and Parks Authority to conduct research. This also corresponds to the design principle: "Bridging between in-school learning and the community" (6).

(v) **Developing a specific curriculum for students for teaching through citizen science:** Once the topic and the scientific needs are defined, the curriculum needs to be developed. This involves determining how many class sessions the curriculum should include, the content, the number of sessions in the field and in the classroom, and student assessment.

Both teachers developed a detailed curriculum and coordinated meetings between experts and the students in the field. Lavi focused on teaching the content of ecology in the field (for example, acquaintance and characterization of biotic and abiotic factors in the grove, characterization of the conditions required for living, acquaintance with biological diversity in general and in the grove in particular, characterization of the grove environment, acquaintance with food chains and food webs—sorting organisms in the grove according to appropriate trophic levels, definition of endangered species, characterization of the rare species in the grove, proposal for ways to conserve nature). During some of the sessions, the students recorded the rare species in the grove to the app.

Hila designed a full annual inquiry-based curriculum with an emphasis on how to identify and formulate a research question, articulate hypotheses, conduct measurements, analyze data, write a research paper, and communicate findings. The content taught included biodiversity, ecosystems and the characteristics of irises, and previous research on irises.

Developing a specific curriculum is an opportunity to implement design principles such as Teaching sequences that are interesting, varied, and clear to the learner (3), Using visual aids to illustrate complex phenomena and ideas (4), Encouraging reflection (5), Creating opportunities for student choice (8), Exposing learners to the breadth of the scientific method (12) and others.

(vi) **Building a framework to assess the contribution of the program**: In every educational program, and especially in an innovative one, evaluation of the program and its contribution to students are a must. In the context of citizen science, the contribution to science is important as well. This evaluation sheds light on ways to improve the program in years to come.

Both teachers researched the program and its impact as part of their master's theses, and as part of work conducted by the Center for Citizen Science at the School. However, even when academic research is not involved, it is desirable to assess the curriculum and its outcomes.

Lavi measured gains in scientific knowledge on pre/post questionnaires. The six items tested knowledge and understanding, applications, and reasoning in the context of ecology. He also compared his students' outcomes to two classes of students from the same school: one similar class of low achievers and a class of high achievers. A one-way ANOVA confirmed that the groups did not differ in extent of knowledge on the pretest. In addition to the scientific knowledge questionnaire, the students filled in a questionnaire on attitudes towards science to assess the level of enjoyment of science classes. Lavi also analyzed the assignments submitted by the students.

Hila evaluated the contribution of participation in the program through participant observations in the classroom and in the field. She collected student reflections during and after the various activities and administered attitude questionnaires towards science before and after the project, as well as a wrap-up reflection questionnaire after the program ended. This enabled Lavi and Hila to determine the impact of their citizen science projects on their students.

(vii) **Implementing the program with flexibility and adaptation:** Despite efforts to design a curriculum carefully, revisions and adaptations need to be made during implementation given the high complexity of this type of open-ended inquiry that includes field work, communication and partnership with scientists and the community, and administrative flexibility on the part of the school.

Lavi had to deal with teaching during the COVID-19 pandemic. The program was delayed until the end of the lockdown (May 2020), so that there were fewer field-based activities. Hila had to deal with changes that occurred in the field. During the study, the class discovered that snails had eaten some of the irises, which meant that the students could not answer their original research question on the flowering of irises and had to change research questions. They also discovered that a plan to incorporate the field into an industrial area was putting the iris at risk. This led to additional content to learn, and activities that were not planned since the students themselves initiated efforts to prevent construction. They published information about the irises and launched a petition, wrote a letter to the mayor, met with local activists, and after realizing that the plan could not be changed, recruited an expert who helped them dig out and replant the irises in a protected area. Thus although it was not intended, science education also shifted toward environmental education, which enabled the students to experience civic action in the form of the protection of nature.

Being able to adjust to such changes is a component of the design principle Cultivating leadership in learners through personal example (9).

These key principles emerged as insights from the bottom-up nature of the two citizen science projects initiated by the teachers. They contributed to the formulation of a series of design principles developed by the TCSS Center to help teachers design citizen science programs, as detailed next.

6.4 Design Principles and Their Implementation in the Programs

This section presents the Design Principles of Citizen Science in Schools developed by the TCSS Center and its community members and demonstrates how they were implemented in the two programs. Of the 27 principles identified by the TCSS community, we present 17 that were relevant to the two programs.

6.5 Discussion and Conclusion

The two projects presented in this chapter show that it is possible to implement successful citizen science projects in schools with diverse student populations, i.e., lower achievers from an under-resourced school as well as higher achievers from a well-resourced school. An ambitious and motivated teacher is pivotal to devising a program that suits the physical environment and the science curriculum, the students, and the management of a project. Planning a co-created citizen science project, from its inception that involves choosing a topic, reconnecting the area, consulting with community members and experts, and preparing a detailed syllabus can last several months. This requires much more from teachers than just pulling a global citizen science off the shelf, such as the project available on Zooniverse, for example. Lavi and Hila, two highly motivated graduate students and exemplary teachers did so with the help of their advisors. The effort required to design and lead a citizen science program from scratch is clearly ambitious and does not suit every teacher. Therefore, we suggest that teachers would benefit from associating themselves with existing well-developed projects. These should not necessarily be off-the-shelf ones. They just need to be structured and provide teachers with scaffolds in terms of learning activities and teacher support. Studies have shown that teachers find it difficult to implement citizen science programs because they are often extra-curricular and do not have associated and appropriate teaching and learning materials, there is little support or professional development so that teachers need to invest additional time and self-learning (Harris et al., 2020). Given their heavy workloads and their obligation to cover the formal curriculum, teachers tend not to pursue citizen science programs (Bonney et al., 2014). Instructors such as Lavi and Hila who initiated and designed their collaborative citizen science programs are rare. In addition, many teachers struggle with outdoor learning and may avoid it due to their lack of confidence in their scientific knowledge, unknown environmental conditions, concerns about student safety, administrative hurdles and others (Kali et al., 2018; Tal & Morag, 2009). However, teachers who find a way to integrate citizen science and outdoor learning in their teaching can successfully motivate their students and develop their nature- and place-attachment, as well as boost and diversify students' learning outcomes. Teachers' interest in meaningful, exciting, and experiential learning that promotes respect for nature, reflect their sense of purpose, autonomy, and high self-efficacy (Evers et al., 2002).

The two projects presented here were successful in meeting their objectives, which included providing their students with authentic learning experiences that allowed them to engage in real-world research that was relevant to their local community. The projects also contributed to scientific knowledge by generating new data and insights. The projects raised awareness of environmental issues in the students who in the case of Hila's 9th-grade class, led to active involvement in the community and activism. Many of the learning outcomes reported in studies that have examined the contribution of participation in citizen science to student learning (Phillips et al., 2018; Lüsse et al., 2022; Peter et al., 2019) also emerged in Hila's and Lavi's findings. This chapter thus contributes to the scant literature on teacher development of a citizen science-based curriculum that serves science teaching in middle schools. The design principles presented here can serve as guidelines for citizen science projects and lay the groundwork for future empirical explorations.

Acknowledgments This research was conducted as part of the TCSS Center of Excellence, which was supported by the Israel Science Foundation (grant no. 2678/17).

References

Ardoin, N. M., Bowers, A. W., & Gaillard, E. (2020). Environmental education outcomes for conservation: A systematic review. *Biological Conservation, 241*, 108224.

Aridor, K., Shaham, R. D., Lavie-Alon, N., Valin, G., Sagy, O., & Kali, Y. (2022). Coalescing knowledge within networks of research practice partnerships: The case of a school-based citizen science network. In *Proceedings of the 16th International Conference of the Learning Sciences-ICLS 2022* (pp. 2114–2115). International Society of the Learning Sciences.

Ballard, H. L., Dixon, C. G., & Harris, E. M. (2017). Youth-focused citizen science: Examining the role of environmental science learning and agency for conservation. *Biological Conservation, 208*, 6575.

Bonney, R., Cooper, C. B., Dickinson, J., Kelling, S., Phillips, T., Rosenberg, K. V., & Shirk, J. (2009). Citizen science: A developing tool for expanding science knowledge and scientific literacy. *Bioscience, 59*(11), 977–984.

Bonney, R., Phillips, T. B., Enck, J., Shirk, J., & Trautmann, N. (2014). *Citizen science and youth education.* National Research Council Committee on Out Of-School Time STEM. National Research Council.

DeWitt, J., & Storksdieck, M. (2008). A short review of school field trips: Key findings from the past and implications for the future. *Visitor Studies, 11*(2), 181–197.

Edwards, R. (2014). The 'citizens' in citizen science projects: Educational and conceptual issues. *International Journal of Science Education, Part B, 4*(4), 376–391.

Evers, W. J., Brouwers, A., & Tomic, W. (2002). Burnout and self-efficacy: A study on teachers' beliefs when implementing an innovative educational system in the Netherlands. *British Journal of Educational Psychology, 72*(2), 227–243.

Haklay, M. (2013). Citizen science and volunteered geographic information: Overview and typology of participation. In D. Sui, S. Elwood, & M. Goodchild (Eds.), *Crowdsourcing geographic knowledge* (pp. 105–122). Springer.

Harris, E. M., Dixon, C. G., Bird, E. B., & Ballard, H. L. (2020). For science and self: Youth interactions with data in community and citizen science. *Journal of the Learning Sciences, 29*(2), 224–263.

Hod, Y., Sagy, O., Kali, Y., Science, T. C., & to School Center, Haifa, Israel. (2018). The opportunities of networks of research-practice partnerships and why CSCL should not give up on large-scale educational change. *International Journal of Computer-Supported Collaborative Learning, 13*, 457–466.

Kali, Y. (2006). Collaborative knowledge-building using the design principles database. *International Journal of Computer-Supported Collaborative Learning, 1*(2), 187–201.

Kali, Y., Levy, K. S., Levin Peled, R., & Tal, T. (2018). Supporting outdoor inquiry learning (SOIL): Teachers as designers of mobile-assisted seamless learning. *British Journal of Educational Technology, 49*(6), 1145–1161.

Lavie Alon, N., & Tal, T. (2017). Field trips to natural environments: How outdoor educators use the physical environment. *International Journal of Science Education, Part B, 7*(3), 237–252.

Lugg, A. (2007). Developing sustainability-literate citizens through outdoor learning: Possibilities for outdoor education in higher education. *Journal of Adventure Education & Outdoor Learning, 7*(2), 97–112.

Lüsse, M., Brockhage, F., Beeken, M., & Pietzner, V. (2022). Citizen science and its potential for science education. *International Journal of Science Education, 44*(7), 1120–1142.

Malone, K. (2008). *Every experience matters: An evidence based research report on the role of learning outside the classroom for children's whole development from birth to eighteen years.* Report commissioned by Farming and Countryside Education for UK Department Children, School and Families.

Oren, L., Tal, T., & Lavie Alon, N. (2022). The contribution of citizen science in an outdoor environment to improving achievements and attitudes towards science of students from low

socio-economic status. In *13th conference of European Researchers in Didactics of Biology (ERIDOB)*. University of Cyprus.
Orion, N. (1993). A model for the development and implementation of field trips as an integral part of the science curriculum. *School Science and Mathematics, 93*(6), 325–331.
Palumbo, R., Manesh, M. F., & Sorrentino, M. (2022). Mapping the state of the art to envision the future of large-scale citizen science projects: An interpretive review. *International Journal of Innovation and Technology Management, 19*(04), 2230001.
Peter, M., Diekötter, T., & Kremer, K. (2019). Participant outcomes of biodiversity citizen science projects: A systematic literature review. *Sustainability, 11*(10), 2780.
Phillips, T., Porticella, N., Constas, M., & Bonney, R. (2018). A framework for articulating and measuring individual learning outcomes from participation in citizen science. *Citizen Science: Theory and Practice, 3*(2), 3.
Pocock, M. J., Tweddle, J. C., Savage, J., Robinson, L. D., & Roy, H. E. (2017). The diversity and evolution of ecological and environmental citizen science. *PLoS One, 12*(4), e0172579.
Rennie, L. J. (2014). Learning science outside of school. In *Handbook of research on science education* (Vol. II, pp. 134–158). Routledge.
Sipos, Y., Battisti, B., & Grimm, K. (2008). Achieving transformative sustainability learning: Engaging head, hands and heart. *International Journal of Sustainability in Higher Education, 9*(1), 68–86.
Tal, T. (2012). Out-of-school: Learning experiences, teaching and students' learning. In B. J. Fraser, K. G. Tobin, & C. J. McRobbie (Eds.), *Second international handbook of science education* (pp. 1109–1122). Springer.
Tal, T., & Morag, O. (2009). Reflective practice as a means for preparing to teach outdoors in an ecological garden. *Journal of Science Teacher Education, 20*(3), 245–262.
Tal, T., Lavie Alon, N., & Morag, O. (2014). Exemplary practices in field trips to natural environments. *Journal of Research in Science Teaching, 51*(4), 430–461.
Tal, T., Barkae, H. S., & Alon, N. L. (2023). From school-based citizen science to transition-driven activism with the community. In J. Diamond & S. Rosenfeld (Eds.), *Amplifying informal science learning: Rethinking research, design, and engagement* (1st ed., pp. 141–154). Routledge.
The National Academies of Sciences Engineering and Medicine. (2018). *Learning through citizen science: Enhancing opportunities by design* (R. Pandya & K. Dibner, Eds.). The National Academies Press.

Chapter 7
The Pedagogical Potential of Citizen Science Projects to Cultivate Scientific Reasoning, Statistical Reasoning, and the Nature of Science Understanding

Haya Ben Simon and Dina Tsybulsky

7.1 Introduction

Life in the twenty-first century is characterized by an overabundance of data and information from a variety of sources. Ordinary citizens need to apply multiple types of reasoning to make informed choices that have a direct impact on their daily lives. People need to think about the data, evaluate the data in light of consensual scientific theories, and have a grasp of the nature of science to do so. In recent years, the involvement of students in citizen science projects has been expanding (Tsybulsky, 2020). The impact of this integration on student skills has not been adequately researched; in particular, the inherent potential of this framework to cultivate reasoning. Studies have shown that citizen science encourages citizen involvement in scientific activities (McKinley et al., 2017) and can help bridge between school culture and scientific culture and between students and scientists (Sagy et al., 2019).

In the case study described in this chapter, the students were involved in a citizen science project. The learning sequence afforded numerous opportunities for statistical investigation and participation in scientific activities that encouraged the students to hone their reasoning skills and learn different aspects of the nature of science (Edmondson et al., 2020). The goal was to better understand when and how citizen science can foster these types of reasoning that are so critical to the decision-making of these future adults.

H. B. Simon (✉) · D. Tsybulsky
Faculty of Education in Science and Technology, Technion Israel Institute of Technology, Haifa, Israel
e-mail: shayak@campus.technion.ac.il; dinatsy@technion.ac.il

D. Tsybulsky, T. Tal (eds.), *Citizen Science Contributions to Biology and Environmental Education*, Contributions from Biology Education Research,
https://doi.org/10.1007/978-3-031-99298-8_7

7.2 Theoretical Framework

7.2.1 *Citizen Science*

Citizen science involves engaging the general public in authentic scientific research. Although citizen involvement varies from project to project, it contributes significantly to many scientific fields including conservation, environmental protection, policy-making, and public engagement (Eitzel et al., 2017). Citizen science is a rigorous research process that provides high-quality data that contributes to problem-solving and advancing science (McKinley et al., 2017). Participating citizens also benefit from involvement in Citizen science projects (Tsybulsky, 2020) since it leads to a better understanding of scientific content and, at times, the scientific process itself. There is increasing evidence of the potential for citizen science to have a significant impact on participants' lives (Bonney et al., 2016).

When students participate in citizen science, they can be involved in projects at varying levels of engagement that are not limited to data collection (Golumbic et al., 2020a, b). In this way, students are exposed to a stimulating scientific experience while learning scientific content (Schuttler et al., 2019). Studies have shown that the framework of citizen science promotes scientific reasoning, statistical reasoning, and an understanding of the nature of science (Aridor et al., 2023; Dvir & Tsybulsky, 2025). Findings show that there is not necessarily a direct correlation between students' level of engagement in citizen science projects and their academic achievements or the educational potential of the projects. Instead, what may have a greater impact is the project design and the development of targeted guidelines for teachers to integrate these projects into their teaching (Berndt & Nitz, 2023).

7.2.2 *Scientific Reasoning*

Scientific reasoning addresses three specific questions about the world: an ontological question—What exists? a causal question—Why is it happening? and an epistemic question—How do we know? (NRC, 2012; Osborne, 2011). The key characteristic of scientific reasoning is the ability to coordinate theory with data arising from scientific inquiry (Gasparatou, 2017). The foundation for this ability is the understanding that scientific knowledge is not only based on experimentation and empirical processes (De Ridder, 2014; Peels, 2017) but also on logic, introspection, and fundamental assumptions. Scientific knowledge involves viewing new data in the light of interpretations and re-interpretations of previous experiments. Data interpretation essentially expresses theories that may be consistent or inconsistent with previously accumulated knowledge. Inconsistency encourages the updating or modification of previous theories. The dynamic process characteristic of science can be summarized as a series of actions, experiences that are either based on or rebuilt, and new accumulated data that are analyzed or reinterpreted to advance explanations and predictions of how the world operates. Scientific paradigms or research programs are constructed to allow various interpretations of the world and to re-evaluate hypotheses. Even the perception of which data can be used to construct a theory is a hypothesis. This perception may change with the development of new scientific theories since different data are considered as evidence in constructing a new theory (Gasparatou, 2017). Most philosophers of science concur that this dynamic process of reconciling data with theory constitutes the core of scientific practices, even if they disagree on how this process unfolds (Laudan & Laudan, 2016). In this context, theory can refer to hypotheses, interpretations, beliefs, or entire belief systems. As for data, it may include a wide range of results, observations, and similar elements. The practice of reconciling data with theory involves a range of reasoning strategies, such as induction or deduction, formulating arguments, including or excluding different types of data, and is influenced by linguistic, historical, and social factors. However, there is a general consensus that several key abilities enable the coordination between data and theory: separating data from theory, proposing alternative interpretations of the same data, and bidirectional alignment; i.e., constructing theory from data and evaluating theory using data (Gasparatou, 2017).

7.2.3 *Statistical Reasoning*

In today's information era, statistics is critical to making informed data-driven decisions (Wild et al., 2018). Statistical reasoning is often viewed as the opposite of the human tendency to draw conclusions from anecdotal evidence. Statistical reasoning involves understanding why and how statistical inquiries are conducted and the underlying principles. These principles include the consistent presence of variability and understanding the characteristics of sampling and the sample. Statistical reasoning is involved in understanding the statistical research process, such as when and how to use appropriate methods for data analysis, how conclusions are drawn from models about larger populations, the role of the context throughout the research process, etc. Statistical reasoning also involves the ability to critique and evaluate the results of statistical research (Garfield et al., 2008).

Statistical reasoning has four dimensions that occur simultaneously during the research process. The first is the research cycle known as PPDAC (the Problem-Plan-Data Analysis-Conclusions inquiry cycle), which involves stating the problem, research planning, data collection and organization, data analysis and hypothesis formulation, drawing conclusions, and going back to the beginning. The second dimension taps general modes of reasoning (e.g., using strategies and tools for problem-solving) and specific modes (e.g., recognizing the importance of data collection). The third dimension is the set of metacognitive skills implemented continuously and recursively, such as interpretation, criticism, and judgment. The fourth dimension is the set of beliefs, approaches, and personal characteristics of the researchers that influence the statistical reasoning process and attitudes towards themselves, statistics, and statistical research, such as skepticism (Wild & Pfannkuch, 1999).

Statistical inference, which is an inseparable part of the first dimension, involves generalization beyond available data by using data as a basic probabilistic language to express uncertainty (Ben-Zvi et al., 2012). Uncertainty is the measure of confidence (or lack thereof) in a developing conclusion. Uncertainty links two concepts: the sample's representativeness (of a larger population) and sampling variability. Novices tend to express an extreme uncertainty or certainty when viewing the sample as either perfectly representative of the entire population (a deterministic view) or as not representing the population at all due to high variability in samples in the same population (a relativistic view). Developing statistical reasoning with uncertainty allows for the emergence of an intermediate area where conclusions can be drawn from the sample but with a certain level of uncertainty (or confidence) (Ben-Zvi et al., 2012). Students' uncertainty is considered to reflect contextual uncertainty or statistical uncertainty (Manor et al., 2014). Contextual uncertainty arises from a conflict between contextual knowledge about the data and the data itself when the data present a different picture than what is known about a certain phenomenon. Statistical uncertainty arises from differences in the data and the limitations of using models of a certain size to conclude about larger populations; i.e., grasping that the sample does not perfectly represent the reality of the entire population.

Fostering statistical reasoning, including the ability to reason with uncertainty and draw statistical inferences, is possible in technological environments that incorporate data analysis (Ben-Zvi et al., 2018; Moore, 1997). Activities involving data research, modeling, and authentic scientific practices contribute to the development and expression of uncertainty reasoning in the intermediate zone (Aridor et al., 2023).

7.2.4 The Nature of Science

Developing students' understanding of the Nature of Science (NOS) takes place through their engagement with the culture of science, and the provision of tools that enable them to critique scientific research and participate in scientific and technological development (Hodson, 2014). These are essential to enable students to participate in public discussions on everyday topics, consume scientific and technological information thoughtfully, continue to learn and scrutinize outside of school (García-Carmona & Acevedo-Díaz, 2018), and choose a suitable career (NRC, 2012).

Many studies have shown that students have naive perceptions about NOS (García-Carmona & Acevedo-Díaz, 2018). Students struggle to grasp the tentative nature of science, the rigor of scientific practice and the explanations it provides for phenomena. Studies show that students, and sometimes even teachers, perceive science as an authoritative, static system that is not subject to change, skepticism, or criticism (Khishfe & Lederman, 2006; Schwartz et al., 2004; Tsybulsky, 2018).

However, instructional pedagogies can foster more informed perceptions. These include pedagogies that emphasize active student engagement in scientific practices and inquiry (Schwartz et al., 2012), research apprenticeships (Burgin & Sadler, 2016), and students' "inquiry" experiences, such as reading adapted scientific literature or visiting university research laboratories (Tsybulsky, 2018, 2019; Tsybulsky et al., 2018a, b). Edmondson et al. (2020) found that learning activities with a high level of authenticity support the development of students' understanding of the NOS. These activities model the entirety of scientific practices engaged in by scientists, from the development of a question to the reporting of results. They also provide an authentic context, whether in the laboratory, field, or other settings like citizen science projects and ensure the authenticity of the findings by contributing genuine knowledge to scientific research.

7.3 Method

This study follows an **instrumental case study approach** (Stake, 1995), in which the case is examined not as an end in itself but as a means to gain insights into broader pedagogical and cognitive processes. Rather than aiming for **statistical generalization**, this study seeks **analytical generalization** (Yin, 2009), where insights derived from the case can inform theoretical understandings of how citizen science projects support scientific reasoning, statistical reasoning, and NOS understanding. This methodological approach allows for an in-depth examination of student engagement in reasoning processes within an authentic learning environment. The findings contribute to the broader discourse on the educational value of citizen science and provide a foundation for future research in diverse educational settings.

7.3.1 Research Context and Participants

Two students aged 15–16 (in the 10th grade) participated in a learning sequence in a citizen science "Sleep" project organized by the TCSS (Taking Citizen Science to School) center, which integrates citizen science projects into schools in Israel. The "sleep project" enables students to experience scientific reasoning, statistical reasoning, and various NOS aspects. The aim was to study the sleeping patterns of Israeli adolescents. The scientists involved in the project were involved in mapping the quality and duration of sleep in Israeli adolescents and probing the potential associations between sleep patterns and features that influence these patterns.

The learning sequence began with a pre-activity where students learned about the scientific background of sleep, including biological clocks, genetics, and their influence on social phenomena. Then, the students were introduced to the scientists' goals and were asked to think about factors that might influence sleep that they would like to collect data about. The students filled out sleep questionnaires documenting their sleep duration and habits. In the second part of the learning sequence, they investigated 14 cases (from weekly sleep questionnaires) and 28 data cases (from bi-weekly sleep questionnaires) using TinkerPlots software. The students were asked to find associations between the characteristics of the data and their own sleep patterns. They were also asked to make predictions about sleep patterns for these characteristics for another period during the year and to formulate hypotheses about the sleep patterns of adolescents in general in the country. The third part focused on constructing models using the sampler tool, which allowed them to graph models based on their hypotheses and then plot simulated samples based on the model. The students built models representing their hypotheses about the associations between a specific characteristic and their sleep patterns. In the final part, students explored Big Data in a file containing data from all the adolescents in the country who participated in the project (5000 cases). They examined their hypotheses in light of these new data and attempted to draw conclusions and new hypotheses.

7.3.2 Data Collection and Analysis

The students' work, videos, and audio recordings of the learning sequence, as well as their transcripts were analyzed using a content analysis approach to identify expressions of scientific reasoning, statistical reasoning, and the NOS in the students' expressions and actions. To assess scientific reasoning, the associations made by the students between theory and data and vice-versa were examined. To assess statistical reasoning, a deductive content analysis identified uncertainties in the students' expressions. To assess the NOS, a deductive content analysis was conducted to identify NOS aspects in the students' expressions using the Family Resemblance Approach (FRA) analytical framework (Erduran & Dagher, 2014).

7.4 Results

7.4.1 Scientific Reasoning

To characterize the students' scientific reasoning, a broad definition of scientific reasoning as a bidirectional adjustment between data and theory was used. Throughout the learning sequence the students frequently referred to theories they were familiar with and sought evidence in the data to support these theories. Conversely, when they encountered other data, they attempted to find explanations that would fit them.

During the first session, the students were asked to think of questions that they wanted to include in the sleep questionnaire for later exploration of the relationship between specific characteristics and sleep patterns. The students raised questions inspired by scientific theories they had heard of: Did I sleep with light on or off? Did I drink caffeine before I went to bed? Did I have a test or a stressful event in the morning?

In the second session, the students examined the data they collected themselves over a 2-week period. They tried to link theories to the data they plotted in the TinkerPlots tool. They also used the word "theory" in their discussions, indicating that they understood the role of theory in providing explanations. For example: "When I had my phone on, the time it took me [to fall asleep] from the moment I decided to sleep was longer because of the radiation." "There is a reasonable explanation for this: the light it emits. When the phone emits white light, it can... deceive the brain, and the brain sees this light as daylight, so it may delay reasoning about going to sleep... But this is a theory." Throughout most of the session, they continued to explore this question and tried to reach a conclusion, but in the end, they decided that more data were needed to support the link between the data and the theory.

Then, the students were asked what they considered to be quality sleep, and in the discussion, they mentioned several different theories they had heard from scientists, related to their data. For example: "I know that, for example, if someone drinks beverages with a lot of sugar or all kinds of foods that cause... like to stay awake, so it may affect [sleep]. And also, heavy meals, I also think they are hard to digest."

The students expressed their hypothesis that someone who drinks caffeine before going to bed is likely to find it hard to fall asleep, and will experience non-continuous and poor-quality sleep. The researchers asked the students if the data could reveal a different picture contradicting this hypothesis. The students replied that this was possible and tried to explain it by expanding on their original theory about the effects of caffeine:

Student A: "There is a chance. It also depends on our age and ... hormones. There is a chance that... now in today's world, we... coffee doesn't... not... help us be more active but rather the opposite. [...]".

Student B: "[...] many things are related to coffee and also it affects people differently ... let's say someone who, who drinks a little bit of coffee every day, it wakes him up in my opinion, and someone who drinks a lot and then suddenly there is a very big difference because he's addicted, so... he can't fall asleep at night. Or he does fall asleep because he is so tired because he didn't drink coffee. So, it's very tricky."

Towards the end of this stage, the students decided to collect additional data on the effects of light in the bedroom: students who are not used to sleeping with the light on would sleep with their light on, and vice versa. The students predicted that someone who sleeps in a darkened room would have more consistent sleep patterns but that a person who sleeps with the light on will have more disrupted sleep patterns, and their characteristics should evidence wider distributions. They based this hypothesis on the data collected from the 28 nights they studied (Fig. 7.1).

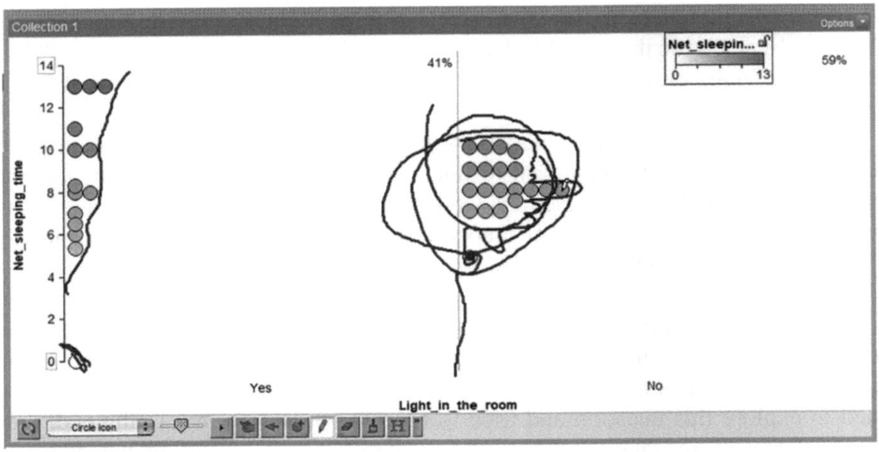

Fig. 7.1 Total Sleep Duration in Light and Dark Conditions. The students marked the distribution of sleep duration when sleeping with light as "extended;" i.e., broader, and the distribution of sleep duration when sleeping in darkness as "narrow and dense;" i.e., more concentrated around certain data points

The students conducted the experiment over a two-night period. The student sleeping with the light on felt that it took her longer to fall asleep and that she woke up more often. However, when they checked these hypotheses against the data, it turned out to be incorrect (Fig. 7.2).

Since their hypotheses were disconfirmed, the students then focused on another question and designed a prediction model using the sampler software: how does (school) pressure affect sleep patterns, particularly the time it takes to fall asleep (Fig. 7.3). In this case, the data confirmed what one student knew from her experience.

Student A: "I know that many times, like, I lie in bed stressed out… and can't fall asleep, so I wanted to check that. So after I calculated the mean, I saw that first, it's narrower, and my routine… when… there's no pressure, then basically the… range of falling asleep is from zero to twenty [minutes] and here [when there's pressure] it spreads out more along the axis."

Then, the students attempted to examine a three-variable axis system consisting of pressure, time to fall asleep, and light on/off (Fig. 7.4a). After examining the data, the students proposed a new theory:

Researcher A: "Let's try to understand what we're seeing. Where are the red dots [representing light condition in the room] located?"

Student B: "They are closer to less time asleep. That means… with light, you sleep for a shorter period of time when you're pressured…".

Student A: "Honestly, I also saw this in the data we collected, that like, even I, when I was pressured, then I felt much more comfortable sleeping with the light on… I wanted to check this hypothesis… the biological effect of light releases things

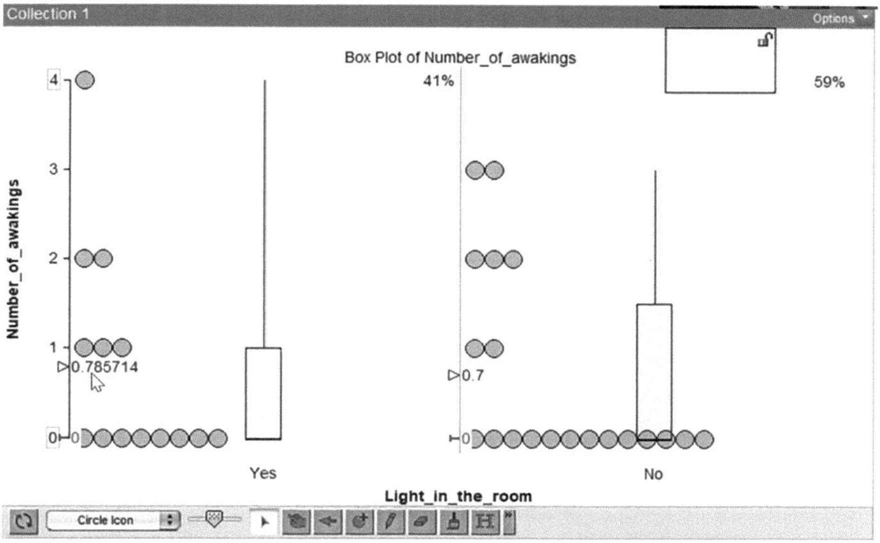

Fig. 7.2 Number of Awakenings in Light and Dark Conditions. The rectangles indicate the range where 50% of the data were located. This range was narrower in the light on condition and more spread out in dark condition, contrary to the students' hypotheses

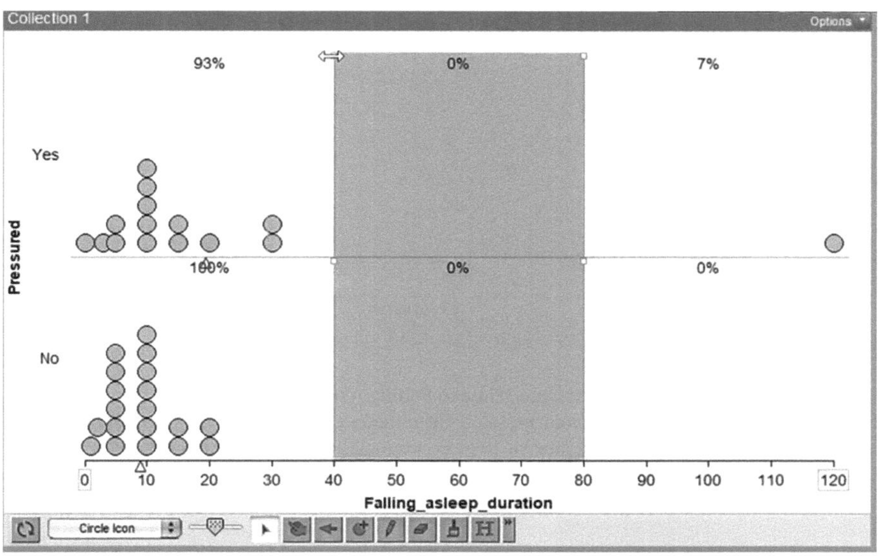

Fig. 7.3 Time to Falling Asleep (in minutes) when feeling pressured before Sleep and without feeling Pressure before Sleep. The data in the pressure category are more dispersed, and the mean time to fall asleep was higher than in the no-pressure category

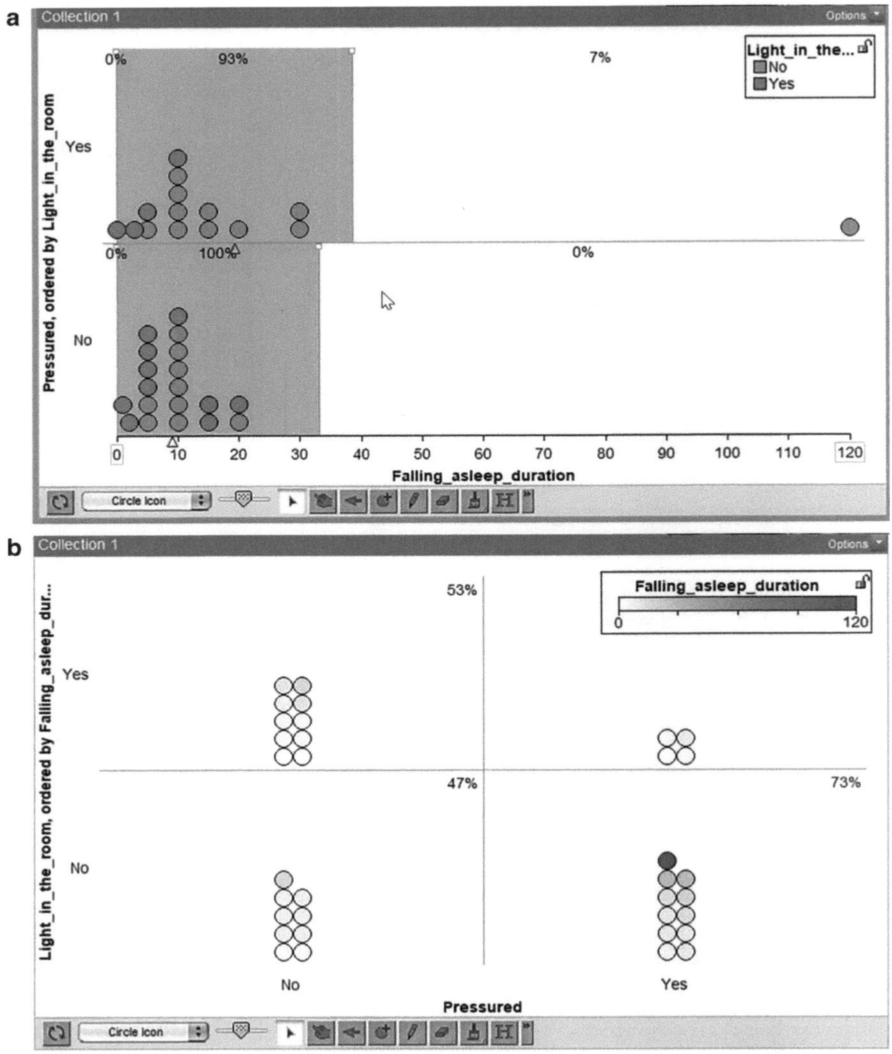

Fig. 7.4 Relationship between Pressure, Time to Falling Asleep, and Light on/off Condition—(**a**) Time to fall asleep with and without pressure (in minutes), in light and dark conditions. Y axis represent pressure when colors represent light conditions. Red indicates light on; blue indicates light off. (**b**) Plot with/without pressure, light on/off, color coded where darker shades indicate longer times of falling asleep duration in minutes

in the brain. And then like, it also involves the pressure, which is interesting because like, like, maybe with your anxiety, you prefer to sleep with light on to be less anxious because the darkness is scary."

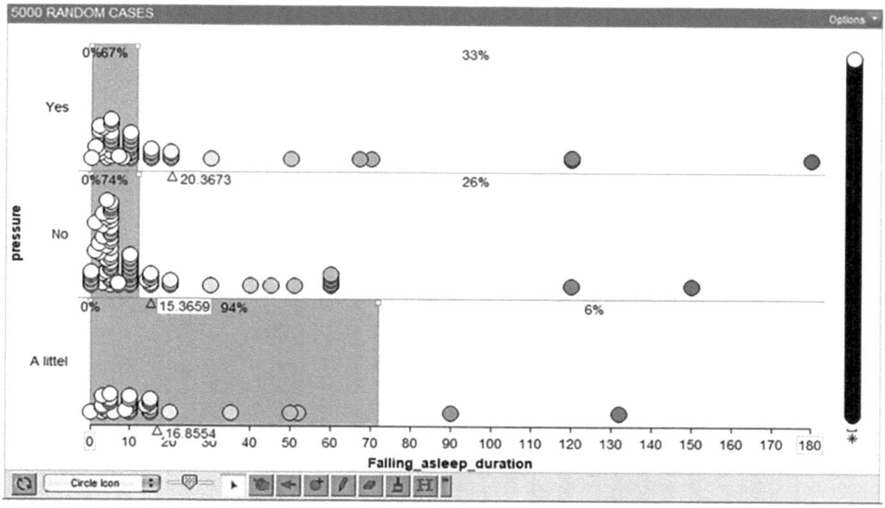

Fig. 7.5 Relationship between Pressure and Time to Falling Asleep (in minutes) in the Big Data. The students plotted the distribution of the data for times ranging from 0–10 min for "Yes" and "No" (in percentages) as well as the means (marked as triangles) and concluded that under pressure, the distribution was wider

The students plotted another graph (Fig. 7.4b) of the data and reached a similar conclusion.

In the fourth part of the learning sequence, the students explored Big Data. The students tested the validity of their hypothesis on the relationship between stress and sleep patterns on a large dataset containing 5000 data cases on adolescent nights. Before seeing the data, Student A expressed some confidence in the hypothesis that the distribution for falling asleep under pressure would be broader. She argued that it made sense for someone experiencing stress to struggle to fall asleep and that there was a biological basis. However, she also expressed an interest in testing the validity of the hypothesis and checking it in a large dataset. The students organized the data in TinkerPlots and generated a suitable graph (Fig. 7.5). They concluded that their hypothesis was correct.

7.4.2 *Statistical Reasoning*

In this section, we chose to present the statistical reasoning in aspect of different types of uncertainties expressed by the students during the learning sequence, as these uncertainties played a major role in their discussions and reasoning.

As early as the second session, the students expressed significant doubts about the reliability of the data, mainly due to the feeling that 2-week data collected from

Table 7.1 Uncertainties expressed during the learning sequence

	Statistical Uncertainty	Contextual Uncertainty	Uncertainty about filling out the questionnaires
Quotes from different points in the learning sequence	"Because it's just one week and let's say for me, what mainly affects the week is what happens on Saturday, let's say what time I go to sleep between Friday and Saturday and like what I'm doing? studying, staying up late." (Investigating 14 data cases) "There is a lot of diversity in the population, and, in my opinion, this sample does not represent, [it can't] represent all adolescents." (Investigating 28 data cases) "Why do you have...only two girls? As if there isn't a group here, for example, of 20 kids where each one of them is a boy, a girl, a boy, a girl, and you can determine the differences." (Investigating 28 data cases) "There are all kinds of things that affect sleep, so... there is too much uncertainty than to say about everyone...the boys... I'm trying to reduce the population so...to control it...so that I can say something about the general, about the population." (Investigating 28 data cases) "I would check again at different periods what happens because there are periods for extreme people like these and then it changes very quickly..." (model generation)	"I didn't feel like I found any connection between the things responses I gave on the questionnaire." (Investigating 14 data cases) "I think it's quite individual, depends on the person, but there's a chance that the body of adolescents with all the hormones and how we grow, caffeine makes us sleep more." (Investigating 28 data cases) "We're two girls... who are in the same school, in the same city, in the same country... similar data and I think that maybe we can think about what it means about us and then see the future for, let's say, the next two months." (Investigating 28 data cases) "You look [at the data] and you need to use your head and analyze it in a way that's not just mathematical but there's a story behind these data, there's something they're telling us, that they're informing us about... to see what it means to each one." (model generation)	"The researcher cannot control... how well the respondent understands the questionnaire... Another researcher who hasn't met the respondents can just do research that is not accurate." (Investigating 28 data cases) "The question is if we're before we [go to sleep] in the evening as if we're going to take a shower, but I showered on the same day but not in the evening and I mark that I didn't shower in the evening but I showered in the morning, I didn't know if it could affect it." (Investigating 28 data cases) "People make mistakes. Every data point for every day may be accurate 70–80 percent. Because people can't be as precise because I wasn't precise about it. How many times did I wake up at night? I woke up and didn't know about it. Like [it's impossible] to know the exact time, yes, the times, the actions..." (Investigating Big Data)

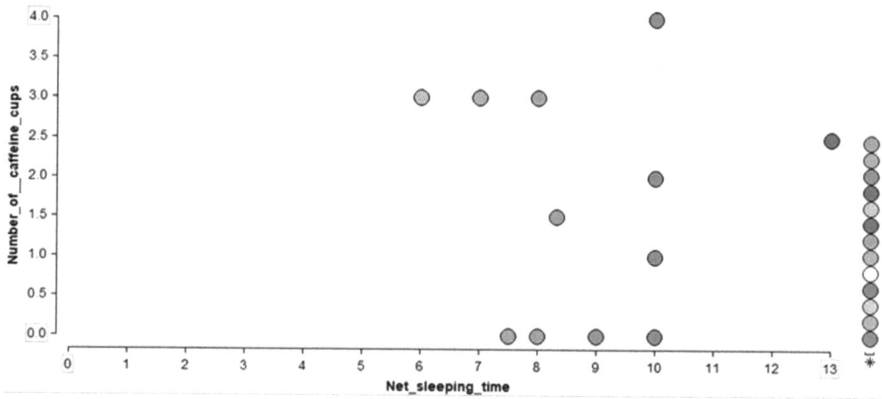

Fig. 7.6 Examination of the relationship between the number of cups of caffeine-containing beverages consumed during the day and the net sleeping time (in hours). A strip of cases along the right side represents the missing data

two students could not provide a broad enough picture. The students realized that some questions were interpreted differently. They asked the researchers about the meaning of certain questions and understood that other students who filled out the questionnaires might not receive similar instructions, thus potentially leading to different answers. The students discussed various types of uncertainties (Table 7.1) and were only willing to make predictions for their own sleeping habits for the next 4 months.

During the discussion and analysis of the different characteristics, one student expressed dissatisfaction with the question "When do you go to bed?" and noted that it may not necessarily reflect the time it takes to fall asleep. However, she acknowledged its importance for measuring healthy sleep. She was willing to trust a larger data sample because she believed that errors in answering the questionnaire would have a smaller relative impact on the larger sample, thus allowing for more reliable conclusions.

Furthermore, when the students explored the relationship between the number of cups of coffee consumed and sleep duration, they noticed that there were a large number of missing data points on the graph (Fig. 7.6). One student understood that the problem was due to the fact that when she did not drink any coffee at all (zero cups), she left the questionnaire line blank instead of writing zero. To address this issue and continue investigating the relationship between these two characteristics, the students decided that all the missing data would be inputted as zero cups of coffee. This example further illustrates how individual students can interpret questionnaire items differently, thus affecting the certainty of the data and the ability to draw conclusions.

During the third part of the learning sequence, the students generated a model based on their hypotheses. One of the models they chose to focus on was the effect of pressure on net sleeping time and time to fall asleep. The students produced one model to test the hypothesis that when a person is under pressure, there should be a wider distribution of time to fall asleep, reflecting the fact that some students fall asleep quickly and others do not. In contrast, when a person is not pressured, the distribution of the time to fall asleep should be narrower and less scattered. Nevertheless, the students expressed statistical and contextual uncertainties. Student A suggested checking other periods and different populations to increase confidence in the hypothesis, while Student B was only willing to test the hypothesis if it referred to the same month.

In the fourth session of the sequence, when the students investigated the Big Data, Student A's confidence increased, and she was willing to formulate hypotheses with less uncertainty. In contrast, Student B continued to express uncertainty despite the high order of magnitude of the data. When asked about the source of her concerns, she replied that despite having a lot of data, neither she nor the other students who filled out the questionnaire provided an accurate picture of the true situation. Several problematic characteristics of the data were raised; for example, the accuracy of the bedtime entered, the decision of when to close their eyes, which may not have been measured at the right time, and the number of awakenings that may not always be recalled correctly. Student A responded that a large amount of data increases confidence, especially when the trend they saw in their sample of 28 data cases repeated itself in the Big Data.

7.4.3 NOS Understanding

In this section presented the major examples of different NOS aspects were expressed in the students' discussions during the learning sequence.

As early as the first meeting, in response to a question about the disadvantages of citizen science, one of the students addressed the importance of precision in science. She expressed concerns that this necessary accuracy would be compromised when the general public, who are not trained and knowledgeable in science, participate in data collection. Later in the same meeting, the students were asked how sleep can be studied. In response, the students listed various scientific methods, including reading publications, studying animals, and conducting surveys.

During the discussions, the students frequently used the word "theory." The teacher asked the students in the next meeting what a theory was and how it differed from a hypothesis. Student A said, "… he [the scientist] had hypotheses and he studies them, and in the end, he finds something that… encompasses everything and has a connection in it… he saw… that his hypotheses were based on. As though they are true. So, he generated something that is like, that is simply his theory."

Student B said, "A theory can become something real when there are facts that justify it. We didn't have, there weren't like facts about our sleep or something until we obtained data that could lead to some result, so a theory will become like the… what happens."

Table 7.2 Students' references to NOS during the learning sequence

Feature of NOS	Example quote	General idea	When expressed within the learning sequence
Shared methods	"You can read about it [sleep] online, talk to someone who has read about it, and learn from them." "You can research sleep in animals." "You can also research people around us... ask about things that interest us."	There is no single scientific method, scientists use a variety of tools	Introduction to the citizen science and Sleep project
Importance of accuracy in science	"I think that sometimes the data collection [in citizen science] can create too many variables... and may lead to errors in the data, and someone who doesn't understand... may not know, for example, that the conditions [in the experiment] need to be identical."	Scientists are accurate in their work so scientific knowledge is based on reliable evidence	Introduction to citizen science and Sleep project, model generation
Reconsidering Knowledge	"Science on these topics is never one hundred percent...we are constantly discovering new things that contradict old ones... every study is based on facts that are true, and that's how science evolves."	Scientific knowledge is relatively stable, but in light of new evidence, it can undergo changes	Investigating 14 cases, Model Generation
Role of Theory	"Even a scientist wouldn't want to research the... hypothesis if he knew for sure what would happen. So, scientists also have situations where their hypothesis is not correct. But they continue to research and find answers that are indeed correct, and then slowly they... they build from such... a collection of... hypotheses, of facts that are theories that can be used."	Theories provide explanations to understand the world	Model Generation
Collaboration among scientists	"If two people conduct research together, for example, they can contribute to each other because each will think differently about things, and they will find more things to prove."	Scientists collaborate in multidisciplinary and multicultural research teams	Model Generation

After this discussion, the students were asked whether theories can change over time, and the discussion revolved around the revision of scientific knowledge and collaboration among scientists. The students believed that the same experiment conducted in different places around the world might lead to different results due to differences in climate and other conditions. They also argued that even two researchers working together in the same place might have different theories related to their different worldviews and diverse perspectives.

One student noted that "every study has a different estimate that can make it inaccurate." In other words, she understood that there are more stringent scientific claims regarding some issues than others. The second student referred to the laws governing the natural world. She understood that a law cannot change, unlike a theory, but struggled to distinguish between the two clearly. From her perspective, a law is an accurate theory that does not change. The student understood that scientific knowledge that is subject to change is valuable and contributes to the construction of theory.

While engaging in model generation, one of the students asked why investigate a question that cannot be answered "100 percent." During the discussion, the students again touched upon the differences between a "law" and a "theory" but did not explicitly differentiate between them. Gravity was mentioned as something that is "surely correct" and will not change, as well as the fact that the world is round and not flat. The student was convinced that there is always value in researching, even when the answer may change over time.

Later in the same conversation, the differences between various scientific disciplines were also discussed, especially between mathematics and biology. Student A commented that, "The difference is that if, say, in biology or psychological research or in… all kinds of medicine and stuff like that, it doesn't always come out the same thing because of all the variables and experiments that are sometimes not the same thing and the… all kinds of changes in society… And mathematics, it's not like that, it's just like, there are laws and according to them, things operate and they discover new things that prove the laws that already exist."

Various NOS aspects discussed by the students during the learning sequence are summarized in Table 7.2.

7.5 Discussion

This study examined the pedagogical potential of a citizen science project to cultivate scientific reasoning, statistical reasoning, and NOS in 10th grade students. The findings indicate that this learning environment allowed for learning experiences and expressions of each type of reasoning.

In terms of scientific reasoning, the results showed that the students were influenced by the theoretical knowledge they were exposed to in the first part of the learning sequence when they learned about the goals of sleep scientists and were told a little about the different issues involved. The students came with a variety of pieces of information that they had heard or read dealing with the associations between certain factors and sleep quality and patterns. This knowledge informed their data exploration process. It influenced the way that the students formulated hypotheses about what they would see in the data, the analysis of the data and its examination, as well as the uncertainty arising from the interpretation of the data. The data often challenged the students' prior knowledge and elicited doubts. As the learning sequence progressed, the students made more attempts to find alternative explanations and more complex explanations for the mismatches they observed between the data and their scientific knowledge. These attempts to coordinate data and theory and theory to data were expressions of the students' scientific reasoning (Gasparatou, 2017). They gradually developed more complex theories that were based on a variety of factors and possible connections, thus showing that they were applying scientific reasoning with the understanding that multiple variables and different mechanisms work together in the biological phenomenon of sleep. Previous studies have shown that students have difficulty constructing scientific explanations, which stems, among other things, from a lack of scientific knowledge (Laliyo et al., 2023). In the current study as well, the students' lack of broader scientific knowledge about the topics or the time to learn about the phenomenon in more depth may have restricted the students' ability to provide deep scientific explanations based on scientific knowledge.

When the students expressed a desire to collect additional data in the light on/light off condition this was another step in the scientific reasoning process. They identified a phenomenon in the data and wanted to examine the variables to ensure that the independent variable (light in this case) affected the dependent variables (time to fall asleep, net sleeping time, and number of awakenings). This is consistent with the literature indicating that scientific inquiry processes with an emphasis on developing scientific reasoning bolster students' ability to identify variables, control them, and analyze the relationships between them (Chen & She, 2015).

In terms of statistical reasoning, the analysis focused on expressions of uncertainty. The students expressed contextual and statistical uncertainties, and well as semantic uncertainty related to the wording of the questionnaires. At the beginning of the learning sequence, when the students examined a small data sample, most concerns revolved around statistical uncertainty related to a sample that is too small to reflect the general population (Manor et al., 2014). Since data were only collected from two students, they raised the question of how well they reflected the entire population contextually, since they did not represent all types of adolescents. As the learning sequence progressed, the students found increasingly more flaws in the questionnaires, making it difficult for them to rely on their conclusions from the data. The problems included inconsistencies in understanding the question and failing to answer a question that was not understood correctly. In the Big Data, the students saw responses from participants who did not provide a suitable answer to the question, and they had to scrub the data to reach insights. Similar difficulties have appeared in previous studies. These include veracity uncertainty, which pertains to the credibility of the data source, and indicates the extent to which students evaluate the trustworthiness of the data source and how it impacts confidence in the data. Research question-data gap uncertainty arises from a discrepancy between the students' initial research inquiry and the actual data provided in the dataset (Gafny & Ben-Zvi, 2023). Similarly, here, the disparity between the students' expectations and the actual data led them to question the integrity of the entire dataset.

In terms of the NOS, the focus on data exploration, particularly in the context of the complex phenomenon of sleep prompted discussions on various aspects of the NOS. As noted by Groth (2018) investigating data on a scientific phenomenon can serve as a gateway to the scientific world, such that the experience of authentic scientific practices in the learning sequence facilitated reflections on different aspects of the scientific enterprise. For example, the concept of "theory" was raised by one of the students, sparking a discussion about the meaning of the term. The students speculated about the accuracy and reliability of the study, which led them to discuss the importance they attributed to these values ??in science. The various discoveries when analyzing the Big Data in the learning sequence also gave the students an experience of tentativeness and prompted discussions about the changes scientific knowledge can undergo as a result of research and collaboration among scientists.

The findings indicated that the framework of citizen science provides a suitable pedagogical framework for students' engagement and cultivation of the three types of reasoning. In line with previous research (Aridor et al., 2023; Berndt & Nitz, 2023; Dvir & Tsybulsky, 2025), this was made possible thanks to the learning sequence that promoted familiarization with the scientific world, data collection, and data exploration.

As this study followed an instrumental case study approach, its contribution lies in offering analytical generalization (Stake, 1995; Yin, 2009) regarding the ways in which citizen science projects can support students' reasoning processes. Rather than aiming for broad empirical generalization, the findings provide theoretical insights into how engagement with citizen science data fosters scientific reasoning, statistical reasoning, and NOS understanding in an authentic learning context.

Further research should focus on the possible interrelationships between types of reasoning and the possibility that the development of a certain form of reasoning encourages the parallel development of another type of reasoning.

Acknowledgments This research was conducted as part of the TCSS Center of Excellence, which was supported by the Israel Science Foundation (grant no. 2678/17). In addition, the research was supported by the Israel Science Foundation (grant no. 1345/22).

References

Aridor, K., Dvir, M., Tsybulsky, D., & Ben-Zvi, D. (2023). Living the DReaM: The interrelations between statistical, scientific and nature of science uncertainty articulations through citizen science. *Instructional Science, 51*, 1–34.

Ben-Zvi, D., Aridor, K., Makar, K., & Bakker, A. (2012). Students' emergent articulations of uncertainty while making informal statistical inferences. *ZDM, 44*(7), 913–925.

Ben-Zvi, D., Makar, K., & Garfield, J. (Eds.). (2018). *International handbook of research in statistics education*. Springer.

Berndt, J., & Nitz, S. (2023). Learning in citizen science: The effects of different participation opportunities on students' knowledge and attitudes. *Sustainability, 15*(16), 12264.

Bonney, R., Phillips, T. B., Ballard, H. L., & Enck, J. W. (2016). Can citizen science enhance public understanding of science? *Public Understanding of Science, 25*(1), 2–16.

Burgin, S. R., & Sadler, T. D. (2016). Learning nature of science concepts through a research apprenticeship program: A comparative study of three approaches. *Journal of Research in Science Teaching, 53*(1), 31–59.

Chen, C. T., & She, H. C. (2015). The effectiveness of scientific inquiry with/without integration of scientific reasoning. *International Journal of Science and Mathematics Education, 13*, 1–20.

De Ridder, J. (2014). Science and scientism in popular science writing. *Social Epistemology Review and Reply Collective, 3*(12), 23–39.

Dvir, M., & Tsybulsky, D. (2025). Facilitating the design and analysis of middle school students' reasoning in the context of citizen science: A framework of the interrelations between statistical, scientific, and nature of science reasoning with data-based claims. *Science and Education, 1–37.*

Edmondson, E., Burgin, S., Tsybulsky, D., & Maeng, J. (2020). Learning about the nature of science through authentic science experience: Realities and potential. In W. F. McComas (Ed.), *The nature of science in science education: Rationales and strategies* (pp. 659–674). Springer.

Eitzel, M. V., Cappadonna, J. L., Santos-Lang, C., Duerr, R. E., Virapongse, A., West, S. E., & Jiang, Q. (2017). Citizen science terminology matters: Exploring key terms. *Citizen Science: Theory and Practice, 2*(1), 1.

Erduran, S., & Dagher, Z. R. (2014). *Reconceptualizing the nature of science for science education*. Springer. https://doi.org/10.1007/978-94-017-9057-4

Gafny, R., & Ben-Zvi, D. (2023). Students' articulations of uncertainty about big data in an integrated modeling approach learning environment. *Teaching Statistics, 45*, S67–S79.

García-Carmona, A., & Acevedo-Díaz, J. A. (2018). The nature of scientific practice and science education. *Science & Education, 27*(5–6), 435–455.

Garfield, J. B., Ben-Zvi, D., Chance, B., Medina, E., Roseth, C., & Zieffler, A. (2008). Learning to reason about samples and sampling. In J. B. Garfield & D. Ben-Zvi (Eds.), *Developing students' statistical reasoning* (pp. 201–214). Springer.

Gasparatou, R. (2017). Scientism and scientific reasoning. *Science & Education, 26*(7–9), 799–812.

Golumbic, Y. N., Baram-Tsabari, A., & Koichu, B. (2020a). Engagement and communication features of scientifically successful citizen science projects. *Environmental Communication, 14*(4), 465–480.

Golumbic, Y. N., Fishbain, B., & Baram-Tsabari, A. (2020b). Science literacy in action: Understanding scientific data presented in a citizen science platform by non-expert adults. *International Journal of Science Education Part B: Communication and Public Engagement, 10*(3), 232–247.

Groth, R. E. (2018). Unpacking implicit disagreements among early childhood standards for statistics and probability. In A. Leavy, M. Meletiou-Mavrotheris, & E. Paparistodemou (Eds.), *Statistics in early childhood and primary education. Early mathematics learning and development* (pp. 149–162). Springer.

Hodson, D. (2014). Learning science, learning about science, doing science: Different goals demand different learning methods. *International Journal of Science Education, 36*(15), 2534–2553.

Khishfe, R., & Lederman, N. (2006). Teaching nature of science within a controversial topic: Integrated versus nonintegrated. *Journal of Research in Science Teaching, 43*(4), 395–418.

Laliyo, L. A. R., Utina, R., Husain, R., Umar, M. K., Katili, M. R., & Panigoro, C. (2023). Evaluating students' ability in constructing scientific explanations on chemical phenomena. *Eurasia Journal of Mathematics, Science and Technology Education, 19*(9), em2328.

Laudan, L., & Laudan, R. (2016). The re-emergence of hyphenated history-and-philosophy-of-science and the testing of theories of scientific change. *Studies in History and Philosophy of Science Part A, 59*, 74–77.

Manor, H., Ben-Zvi, D., & Aridor, K. (2014). Students' reasoning about uncertainty while making informal statistical inference in an "Integrated Pedagogic Approach". In: K. Makar, B. de Sousa, and R. Gould (Eds.), Sustainability in statistics education (Proceedings of the Ninth

International Conference on Teaching Statistics ICOTS9 July 2014), International Association for Statistical Education and International Statistical Institute, Voorburg.
McKinley, D. C., Miller-Rushing, A., Ballard, H. L., Bonney, R., Brown, H., Cook-Patton, S. C., Evans, D. M., French, R. A., Parrish, J. K., Phillips, T., Ryan, S. F., Shanley, L., Shirk, J., Stepenuck, K. F., Weltzin, J. F., Wiggins, A., Boyle, O. D., Briggs, R. D., Chapin, S. F., Hawitt, D. A., Preuss, P. W., & Soukup, M. (2017). Citizen science can improve conservation science, natural resource management, and environmental protection. *Biological Conservation, 208*, 15–28. https://doi.org/10.1016/j.biocon.2016.05.015
Moore, D. S. (1997). New pedagogy and new content: The case of statistics. *International Statistical Review, 65*, 123–137.
National Research Council. (2012). *A framework for K-12 science education: Practices, crosscutting concepts, and core ideas*. National Academies.
Osborne, J. F. (2011). Science teaching methods: A rationale for practices. *School Science Review, 93*(343), 93–103.
Peels, R. (2017). Ten reasons to embrace scientism. *Studies in History and Philosophy of Science, 1*(63), 11–21.
Sagy, O., Golumbic, Y. N., Ben-Horin Abramsky, H., Benichou, M., Atias, O. M., Braham, H., Baram-Tsabari, A., Kali, Y., Ben-Zvi, D., Hod, Y., & Angel, D. (2019). Citizen science: An opportunity for learning in the networked society. In Y. Kali, A. Baram-Tsabari, & A. Schejter (Eds.), *Learning in a networked society* (pp. 97–115). Springer.
Schuttler, S. G., Sears, R. S., Orendain, I., Khot, R., Rubenstein, D., Rubenstein, N., Dunn, R. R., Baird, E., Kandros, K., O'Brien, T., & Kays, R. (2019). Citizen science in schools: Students collect valuable mammal data for science, conservation, and community engagement. *Bioscience, 69*(1), 69–79.
Schwartz, S., Lederman, G., & Crawford, A. (2004). Developing views of nature of science in an authentic context: An explicit approach to bridging the gap between nature of science and scientific inquiry. *Science Education, 88*, 610–645.
Schwartz, R. S., Lederman, N. G., & Abd-el-Khalick, F. (2012). A series of misrepresentations: A response to Allchin's whole approach to assessing nature of science understandings. *Science Education, 96*(4), 685–692.
Stake, R. E. (1995). *The art of case study research*. Sage.
Tsybulsky, D. (2018). Comparing the impact of two science-as-inquiry methods on the NOS understanding of high-school biology students. *Science & Education, 27*(7), 661–683. https://doi.org/10.1007/s11191-018-0001-0
Tsybulsky, D. (2019). Students meet authentic science: The valence and foci of experiences reported by high-school biology students regarding their participation in a science outreach programme. *International Journal of Science Education, 41*(5), 567–585. https://doi.org/10.1080/09500693.2019.1570380
Tsybulsky, D. (2020). Self-reported reasons for participating in pro-environmental citizen science activities: A case study of butterfly monitoring in Israel. *Frontiers in Education (section STEM Education), 5*(116), 1–10. https://doi.org/3389/feduc.2020.00116
Tsybulsky, D., Dodick, J., & Camhi, J. (2018a). The effect of field trips to university research labs on Israeli high school students' NOS understanding. *Research in Science Education, 48*(6), 1247–1272. https://doi.org/10.1007/s11165-016-9601-3
Tsybulsky, D., Dodick, J., & Camhi, J. (2018b). High-school students in university research labs? Implementing an outreach model based on the 'science as inquiry' approach. *Journal of Biological Education, 52*(4), 415–428. https://doi.org/10.1080/00219266.2017.1403360
Wild, C. J., & Pfannkuch, M. (1999). Statistical reasoning in empirical inquiry (with discussion). *International Statistical Review, 67*, 223–265.
Wild, C., Utts, J., & Horton, N. (2018). What is statistics? In D. Ben-Zvi, K. Makar, & J. Garfield (Eds.), *International handbook of research in statistics education* (pp. 105–137). Springer.
Yin, R. (2009). *Case study research: Design and methods* (4th ed.). Sage.

Chapter 8
What Do Teachers Need to Bring Citizen Science into the Classroom? Lessons Learned from the Snuffelfiets Project

Bart de Groot and Jelle Boeve-de Pauw

8.1 Introduction

Citizen science initiatives are "projects in which volunteers partner with scientists to answer real-world questions" (Bonney et al., 2014). Through this collaboration with citizens, genuine scientific data and knowledge are generated (Robinson et al., 2018). This does not only allow for acquiring data through non-traditional means, but it can also be used to establish wide public participation, potentially leading to an increase in scientific knowledge and positive attitude towards science throughout society (Bonney et al., 2009). Not only can citizen science be beneficial for scientists and society, it could also be used as a tool for learning in formal education. Through participation in genuine research, citizen science could be used to engage students by connecting science directly to their lives (Jenkins, 2011). It also lends itself to integrative and interdisciplinary practices, connecting the natural, physical, and health sciences to the humanities and social sciences (Pykett et al., 2020), making it an interesting tool for biology and environmental education.

Teachers could implement citizen science projects as an introduction to the scientific process or to achieve a specific learning objective. Research into learning outcomes of citizen science projects shows that it can lead to increased understanding of the nature of science and topic-specific, science-specific, and generic knowledge (Aristeidou & Herodotou, 2020). Harlin et al. (2018) argue that similar outcomes might be achieved in citizen science projects in formal education.

However, whether teachers are able to reach the additional educational goals they have set for the project or are only able to collect data for the scientists largely, depends on how they implement the project. Citizen science projects often have

B. de Groot · J. Boeve-de Pauw (✉)
Freudenthal Institute, Faculty of Science, Utrecht University, Utrecht, Netherlands
e-mail: j.n.a.boeve-depauw@uu.nl

D. Tsybulsky, T. Tal (eds.), *Citizen Science Contributions to Biology and Environmental Education*, Contributions from Biology Education Research,
https://doi.org/10.1007/978-3-031-99298-8_8

some form of facilitation to make sure the project progresses smoothly. Sometimes the citizen scientists are in contact with the researchers and are directly facilitated by them. Other times facilitators take care of the on-site facilitation of the project in the case that researchers are not up to the task (on their own) or are simply absent.

In formal education, teachers often are the facilitators of the project for their students. Unlike facilitators who facilitate a project (solely) on behalf of the researchers, teachers face a problem particular to formal education, as they have to integrate the project into their teaching. They often do not feel supported in adapting such projects to the context of formal education (Aristeidou et al., 2022). They struggle with making time to organise the activities, developing methods and instruments, finding resources, and maintaining their own and their students' motivation. Roche et al. (2020) add other challenges, like constraints set by policies, (felt) insufficiencies in an individual teacher's skillset, and the absence of a digital or communicative infrastructure. This lack of support can in turn have negative effects on how students engage with the project (UC Davis Center for Community and Citizen Science, 2018).

Several ideas have been put forth to make it easier for teachers to implement citizen science projects into their teaching. In the design phase of the project, it could be directly tied to the curriculum or connected to the local community of the school (Harlin et al., 2018). However, these design claims are based on the experiences of a teacher-created citizen science project, in which teachers design their own project relevant to their local environment in conjunction with scientists. These co-created projects are more likely to achieve educational and scientific goals, but they are highly labour intensive to develop for all parties (Gray et al., 2012). Not all teachers have the chance to commit a lot of time to the development of such projects, so it might prove more fruitful to look into claims focused on the facilitation of projects.

Another approach is to train teachers in the facilitation of citizen science projects available to the public. In the research of Lorke et al. (2019), a list of training needs for project designers, facilitators, and participants of citizen science was put forward. Their definition of a facilitator of citizen science included those who train and guide participants of a citizen science project. In the context of formal education this role falls on teachers. Because the facilitator category is quite broad, the determined training needs are, although relevant for the overall facilitation of citizen science, not specified for individual professions. Whether and to what extent they apply to the experience of teachers in particular remains to be researched.

Therefore, the aim of this research is to explore how (science) teachers wish to be supported in adapting and integrating citizen science projects into their teaching. This is done through interviews with teachers who have participated in the Snuffelfiets project with their students using the training needs identified by Lorke et al. (2019) as a leitmotif.

8.2 Background

8.2.1 Citizen Science

Citizen science is science that allows non-scientists to actively participate in scientific research. In projects set up by scientists themselves or organisations for the advancement of science, a large network of amateur or non-professional scientists will gather, analyse, and interpret data for scientific research. In this manner even citizens who have little to no experience in the field of research can contribute to science. For certain scientists it is an interesting opportunity, because it "offers a means of doing substantial, thoughtful public outreach, and of tackling otherwise intractable, laborious, or costly research problems" (Gura, 2013, p. 259).

Citizen science has been around for quite some time. The term was recorded for the first time in the research of an article placed in the *MIT Technology Review* issue of January 1989, which involved 225 participants who collected and inspected rain samples across all 50 states of the USA (Haklay et al., 2021; Kerson, 1989). However, even before that, examples of citizens participating in a form of science or research can be found throughout history (Aono & Kazui, 2008; Finkel, 2018). It is difficult to gauge how much citizen science has grown since, but it cannot be denied that projects are becoming ever more common (Pelacho et al., 2021).

In the 2010s, there has been a significant increase in the number of academic articles published on citizen science, a trend that coincides with several digital citizen science projects that made use of a web-platform to reach out to non-scientists (Kullenberg & Kasperowski, 2016). Web-platforms such as The Citizen Science Alliance, SciStarter, and Citizen Science Central have several hundreds of citizen science projects available (Gura, 2013). These organisations also include Dutch projects, but there are Dutch organisations that provide a similar service as well, such as the RIVM, the Dutch National Institute for Public Health and the Environment (Dutch: Rijksinstituut voor Volksgezondheid en Milieu) (RIVM, n.d.), and GlobeNL, the Dutch branch of the international organisation GLOBE which brings citizen science projects to schools and students worldwide (GlobeNL, 2023).

What exactly constitutes a citizen science project is still a matter of debate (Haklay et al., 2021). The concept has proven to be difficult to define, as it depends on the context in which it is presented. In addition, a broad range of projects across different disciplines have been assigned the denomination. For example, within the natural sciences, citizens could monitor biodiversity all over the world (Schmeller et al., 2009) or measure temporal dynamics of aerosol distribution (Snik et al., 2014), but there are also citizen science projects in the social sciences and humanities (Tauginienė et al., 2020).

8.2.2 The Role of Teachers in Citizen Science

Because teachers are the facilitators of the citizen science project, their involvement is key to the success of the project. There are many challenges that the teachers may face while implementing the project (Aristeidou et al., 2022; Roche et al., 2020), and the many new areas to explore might place the teachers in an unusual, challenging role as learner (Fazio & Karrow, 2015, as cited in Aristeidou et al., 2022). How teachers deal with these challenges seems to be affected by their confidence in their own scientific literacy (Jenkins, 2011), and by their subject-specific training and the desire to develop themselves further (Kelemen-Finan & Dedova, 2014). Therefore, Aristeidou et al. (2022) recommend teachers (and other facilitators) be trained for their role in projects like this.

8.2.3 The Framework

In their report on the training needs for those involved in citizen science, Lorke et al. (2019) established a framework of (training) needs for the different audiences of citizen science, i.e. participants, facilitators, and designers. They identify that facilitators are people who train, educate, or lead participants in a citizen project. This is a role that fits the teachers who have guided their students in the Snuffelfiets project. The needs of these audiences are divided into three broad categories: core needs, operational needs, and engagement needs. According to Lorke's framework, the exact needs within each category differ between the different audiences, but the overall content remains the same:

> The 'Core' needs are those related to the holistic and epistemological nature of the project. […] The 'Operational' needs are concerned with training related to the project processes. […] The 'Engagement' needs [are] concerned with training related to the experiential needs of those involved in the project. (Lorke et al., 2019)

8.2.4 The Snuffelfiets Project

The participants of this study are all teachers with students who have participated in one particular environmental research project: the Snuffelfiets project. In this project, participants are asked to place a sensor (see Fig. 8.1) on their bicycle, measuring the amount of particulate matter in the air while they ride around their neighbourhoods. Through this, data is gathered for scientific research done by the RIVM. The data is also openly accessible through websites like knowyourair.net (SODAQ, n.d.), see Fig. 8.2. This project is currently facilitated by the citizen science organisation GlobeNL (2023).

The reach of the project is broader than just schools and students, but GlobeNL is focused on this particular audience. They provide schools with teaching material

Fig. 8.1 The sensor as used in the Snuffelfiets project. (Note. Reproduced from *Kans voor nieuwe scholen om mee te doen met de snuffelfiets* by GlobeNL 2022. (https://globenederland.nl/onderzoeksprojecten/fijnstof/kans-voor-nieuwe-scholen-provincie-utrecht-snuffelfiets/), Copyright n.d., SODAQ)

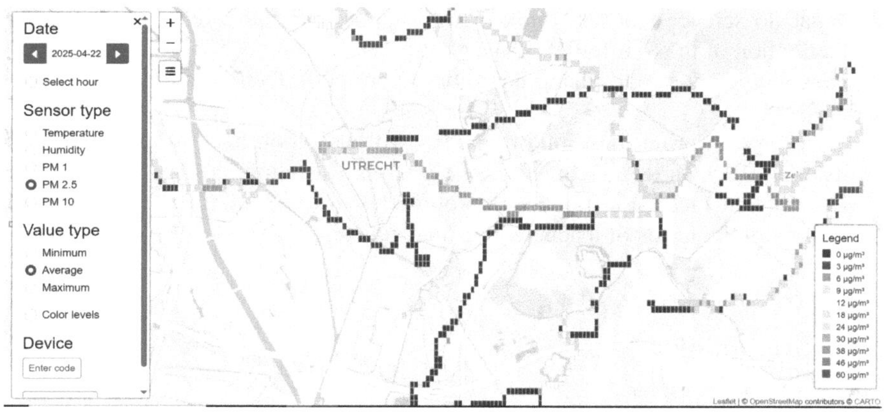

Fig. 8.2 Example of the knowyourair.net website from June 16th, 2023. (Note. The colour of each square represents the amount of particulate matter in the air as measured by the Snuffelfiets sensor. Reproduced from *KnowYourAir* by SODAQ, n.d. (https://knowyourair.net/). Copyright n.d. SODAQ)

and support so that teachers can implement the project (GlobeNL, 2023). The project lends itself to extracurricular activities and classroom projects, and because of the environmental research subject matter, it is possible to implement it not only in biology and other science subjects, but also in geography, research & development (Dutch: Onderzoeken & Ontwerpen or abbreviated O&O), and other interdisciplinary subjects.

One such subject is Nature, Life, and Technology (Dutch: Natuur, Leven, en Techniek or abbreviated NLT), a non-standardised subject with a modular format (NLT, n.d.). Schools that are subscribed to the NLT program can choose among tens of modular learning units to teach their classes. One of these modular units is directly connected with the Snuffelfiets project and expands further upon the teaching material already freely available.

8.3 Research Question

To explore the needs of Dutch science teachers in implementing citizen science projects into their teaching, the main research question is:

> *Which needs do science teachers in Dutch secondary education experience in the implementation of citizen science projects into their teaching?*

Guided by the framework of Lorke et al. (2019), we will answer this question from the specific viewpoint of the Snuffelfiets project and focus on those elements that are specific to the project as well as those that are relevant for citizen science projects in general.

The research question is split up into the following sub-questions:

1. In what ways are science teachers supported in their role as facilitator of the Snuffelfiets project?
2. What do science teachers think of the support they have received during the facilitation of the Snuffelfiets project?
3. How would science teachers like to be supported in facilitating the Snuffelfiets project?

   After answering these questions, we will also look back at the framework itself and use our data to suggest answers to the question:
4. How can the facilitator framework proposed by Lorke et al. (2019) be adapted to better suit the needs of teachers in formal education?

8.4 Methodology

This study applies a qualitative approach to answering the above research questions. Semi-structured interviews have been performed with teachers in secondary education who have implemented the Snuffelfiets project in their classroom. Semi-structured interviews were chosen because all elements of the framework needed to be addressed, but this method still allows the interviewees to expand upon their thoughts.

8.4.1 Participants

Seven teachers were interviewed, coming from seven different schools in two provinces in the Netherlands (see Table 8.1). They all participated in the same Snuffelfiets project, but there were several different ways that the project had been implemented, e.g. mandatory participation in class and voluntary participation in extracurricular activities. The participants ranged in teaching experience and scientific background. In line with the Utrecht Faculty of Science Ethics Quick Scan, all participants gave active consent to be interviewed, recorded, and included in the research.

Table 8.1 Participant demographics

Participant	Subject	Teaching Experience (Years)	Province	Student Participation
Teacher 1	Physics	7	Utrecht	Voluntary
Teacher 2	Physics	16	Utrecht	Voluntary
Teacher 3	Physics/NLT	19	Utrecht	Mandatory
Teacher 4	Physics/NLT/Science	10	Utrecht	Mandatory/ Voluntary
Teacher 5	Research & Development (O&O)	2	Noord-Holland	Mandatory
Teacher 6	Chemistry/NLT/Science	20	Noord-Holland	Mandatory
Teacher 7	Science	10[a]	Noord-Holland	Mandatory

[a]As technical education assistant, teacher in training

8.4.2 *Interviews*

The interview scheme for the semi-structured interviews was based on the report by Lorke et al. (2019). The training needs of facilitators have been translated into several interview questions, divided into three different categories: core needs, operational needs, and engagement needs. As an example, the need for a scientific background of the project was translated to "Was the scientific background of the project clearly explained?" with follow-up questions regarding the relevance of this information, whether this information served as material to expand upon, etc. The interviews were semi-structured in nature to address all topics but still allow the participants to explain their answers (Denscombe, 2010). After the first interview, which also functioned as the pilot, the interview scheme was updated to address prompting the interviewee to expand upon their answer.

8.4.3 *Data Collection, Analysis, and Interpretation*

Interviews took place online and were recorded. Each individual interview took between 45 min to an hour. The interviews were transcribed into Dutch automatically by Microsoft Teams. After the interviews, the transcriptions went through multiple rounds of coding. A combination of a deductive and inductive approach was used. In the first round, mentions of particular elements of the framework of Lorke et al. (2019) were coded in a deductive manner. Thereafter, sub-codes were inductively added to each element along three dimensions: how the particular element was currently implemented in the project, the opinion of the teacher on how the element was implemented, and suggestions from the teacher on how to improve the element. In a third round, codes and sub-codes for elements that did not fit the needs of Lorke et al. (2019) were inductively added.

8.5 Results and Discussion

In this section we will discuss the answers to the interviews. Due to the amount and interconnectedness of the answers, we have elected to summarise the discussion of the interviews along four overarching themes. For a more detailed elaboration of the answers to the interview questions per theme of the Lorke et al. (2019) framework, look at **Appendix C**. A summary can be found in Table 8.2 in the form of suggested changes to the framework.

8.5.1 Facilitating Role

Because the participating teachers already have experience teaching students, the role of facilitator came naturally to them. To varying degrees, they were confident in their own ability to lead their students in the project, and they were able to provide them room for learning largely on their own. Lorke et al. (2019) already are aware that facilitators can bring their own expertise to a citizen science project and want to allow them time to give their feedback.

However, this does not mean that teachers do not want to be supported at all in the implementation. The teachers appreciate the general information given on the subject matter, as they, just like their students, are often new to the subject. For most participating teachers, this information did not need to be highly elaborate, as they teach young students who only have limited experience in and knowledge of the sciences, and they often do not have the time to go in depth into the subject. It just needs to be enough for the students so that they "[know] what they [are] working on and what they [are] talking about."

Also, help with facilitation of the project protocol was welcome. The participating teachers were happy with the instructions they had received during meetings with GlobeNL and through instruction manuals. This helped them quickly explain the way the data should be gathered with the sensors and could be accessed online. Extra help with navigating and processing some of the data, however, would have been appreciated, as the created data sheets were difficult to read for both students and teachers who are inexperienced with programs like Excel. Setup of the sensors also required assistance, as due to how the internet was accessed in some schools, the filters needed to be bypassed. This required outside help from the IT department within the school.

The participating teachers appreciated the teaching materials given to them by GlobeNL, especially those related to doing research. Not every teacher will have the same level of experience with research, and some will need help with, for example, teaching about good research questions and practices. The inclusion of the research cycle as viewed by GlobeNL was a great addition, as its representation and interpretation could differ between groups. Making it explicit for the teachers helped them

Table 8.2 The facilitator framework according to Lorke et al. (2019) and the revised teacher framework

Category	Facilitator Needs (Lorke et al., 2019)	Teacher Needs (Study Results)
Core Needs	**Definition of citizen science—**The nature of citizen science needs to be understood as a collaboration between individuals and the scientific community	**Definition of citizen science**—*Teachers need a clear definition of citizen science and connection with actual research so that it motivates students to work on the project*
	Scientific background—The scientific background, concepts, and skills need to be clear enough that they give the facilitators the confidence to pass the information on to the participants	**Scientific background**—*Teachers facilitate the project for students, so in most cases the scientific background does not need to go highly in depth. Scientific research could be used as a motivator as well, so provide teachers with examples and explain how they could make use of them*
	Relevance—Facilitators need to be aware of the local and global relevance of the project to explain it to their participants, and they need to be aware of curricular and pedagogical concepts relevant to the project	**Relevance**—Teachers need to be aware of the local and global relevance of the project, *especially because relevance serves as a motivator for their students. Depending on the way a project is implemented, curricular and pedagogical relevance may or may not need to be addressed in depth*
Operational Needs	**Project design—**Facilitators need to be aware of the project design and its accurate implementation so they can confidently communicate the practical aspects to their participants, and facilitators need to be comfortable with the project design	**Project design**—Teachers need to be aware of the project design and its accurate implementation so they can confidently communicate the practical aspects to their students. *If possible, allow teachers freedom in the implementation, but guide them around possibilities*
	Project protocols—Facilitators need to be aware of the practice of project protocols and the use of any technical or scientific equipment used in the project so they can confidently communicate the practical aspects to their participants	**Project protocols**—Teachers need to be aware of the practice of project protocols and the use of any technical or scientific equipment used in the project so they can confidently communicate the practical aspects to their students. *Special attention may need to be given to setting up equipment in a protected school environment*
	Project platform—Facilitators need to be aware of how to make use of the project platform, e.g. for communication and data management, so they can confidently communicate the practical aspects to their participants	**Project platform**—Teachers need to be aware of how to make use of the project platform, e.g. for communication and data management, so they can confidently communicate the practical aspects to their students
	Additional material—Facilitators need to be aware of what extra material is available to them and how they could use it in their facilitation of the project	**Additional material**—Teachers need to be aware of what extra teaching materials are available. *If possible, teachers should have access to all additional material for inspiration. As teachers have plenty of experience teaching, they know how to rewrite material to fit their own students, so provide them with adaptable material*

(continued)

Table 8.2 (continued)

Category	Facilitator Needs (Lorke et al., 2019)	Teacher Needs (Study Results)
Engagement Needs	**Recruitment—**If applicable to the project, facilitators need to know how to recruit participants	**Recruitment**—*Teachers can implement a citizen science project into their teaching in different ways, so teach them about recruitment approaches for mandatory and voluntary participation, targeted at students*
	Enjoyment—Facilitators need to ensure that the project is meaningful, engaging and enjoyable for the participants	**Enjoyment**—*Teachers have experience working with students and rely on that to facilitate enjoyment. However, particular projects or topics can inherently have positive or negative effects on enjoyment, so support may still be needed*
	Communication—Facilitators need to be aware of how to facilitate communication within the project, e.g. between participants themselves or between participants and the general community. They also need to be aware of how to best make use of the communication channels available to them	**Communication**—*Teachers have experience working with students and rely on that to facilitate communication. However, help with and activities around communication are still appreciated. Provide teachers with communication channels with other teachers for sharing ideas and feedback*
	Recognition and reward— Facilitators need to be aware of opportunities to recognise and reward their participants, and they need to be provided with ways to do so	**Reward and recognition**—*Teachers highly value feedback for their students' learning, especially feedback coming from researchers and others involved with the topic outside of a school environment. Provide teachers with and make them aware of ways to give their students reward and recognition*
	Wider involvement—Address opportunities for facilitators to become more involved in the citizen science projects	**Wider involvement**—*Teachers enjoy the opportunity to give feedback to the project group. Allow them opportunities to give their opinion on the facilitation of the project and value their input. Involvement in other citizen science projects can be interesting for teachers, but time may be an issue*
Organisational Needs		**Management**—*Teachers need to be trained in handling citizen science projects effectively in a school environment. Help them set up a framework or team, or help them train a coordinator, which in turn could help them implement citizen science projects successfully in their school*

Note. In the right column the results of this study are summarised along the framework of Lorke et al. (2019). The proposed changes to the framework are given in italics. In the left column the original needs of the facilitator framework are represented for comparison

to recognise what doing research in the project entailed for GlobeNL, which helped them to facilitate the project for their students and reward their progress.

However, not all participating teachers made use of the available material. This depended on the strictness of the research protocol, as participating teachers who allowed their students more freedom to do their own research were the ones who stated they did not make much use of the resources. This was because the material "did not fit [the plans of the students] completely." It is therefore good to keep in mind that accompanying material fits the level of strictness of the protocol. On the other hand, some participating teachers were willing to rewrite or repurpose the given material to fit their and their students' needs despite time constraints. Teachers have experience deciding what level is appropriate for their students and know what is feasible to do within the timeframe they have with them. It is therefore appropriate to "include adaptable lesson plans to support teachers creating their own lesson plans" (Harlin et al., 2018, p. 425).

Several participants would have liked to have had a look at more of the material that was made for the project. Part of the material was hidden behind paid membership to the NLT subject. Non-NLT teachers would have liked to have access to at least some form of the information within the NLT module to see whether it would fit their class or they could otherwise make use of it.

One other thing that should not be discounted is the importance of a supporting group. Although the participating teachers did fall back a lot on their own teaching experience, all have stated that they enjoyed the communication between teachers, and the ability to exchange experiences and tips was seen as valuable. This is an element not yet included in the training needs of Lorke et al. (2019), but Harlin et al. (2018) and Aristeidou et al. (2022) suggest giving teachers opportunities for networking with others involved in the project.

8.5.2 Student Participants

Another aspect to keep in mind is that teachers are facilitating the citizen science project for student participants. When the participating teachers broached the subject of citizen science to their students, it was often the first time that these students were coming into contact with it. In addition, it was the students who were approached by the teachers to participate in the project, mandatorily or voluntarily. They differed from participants who come into contact with citizen science completely through their own interests. Students therefore require a different approach for recruitment and engagement than a general audience.

Many of the participating teachers placed an emphasis on motivating their students, similar to the results of a survey in the research of Aristeidou et al. (2022). They expressed that multiple elements of the project were or could have been great motivators for their students. For example, the chance to measure air quality in their own neighbourhood was seen as especially motivating, as one teacher said that "bringing their route from home to school into view, [...] students [were] quickly inclined to ask questions." What exactly the participating teachers wanted to use the motivators for differed slightly between participants who made participation in the project voluntary and mandatory.

Teachers who made student participation in the project voluntary encountered more difficulties while recruiting students. However, as soon as they were recruited, they did not encounter any problems with keeping their students engaged. Teachers who made student participation in the project mandatory did not run into difficulties recruiting their students, but it was more difficult to keep them engaged. It is important to keep in mind the different ways a project could be implemented in a school and help guide teachers accordingly.

Depending on the subject matter and/or the method of research (e.g. the way data is gathered), the project can be inherently (de)motivating for the students. In the case of particulate matter, the participating teachers believe that the subject was, at first, not very appealing as it did not concern their students' daily lives. Doing the research itself, on the other hand, did appeal to the students, as the project allowed them to perform their own research in their own neighbourhoods. It is, however, important to keep them engaged beyond doing their own research, as it was noticed that student engagement with the project decreased once the students had completed their own work.

In general, it appears to be important for the participating teachers to have a clear idea of how they will implement the project with their students before they start. This is especially the case for teachers who recruit students on a voluntary basis, as it would help them pitch the project. During the Snuffelfiets project, this led to some trouble for some participating teachers, as "when [they] stepped into the project, [they] didn't have a clear vision of how to properly implement it in [their] education." According to them, this caused difficulties when recruiting students. It is therefore recommended to implement general guidelines for the project if possible, making it easier for the teachers to create a project guideline that fits their needs and ideas for the project.

8.5.3 The Bigger Picture

The third aspect that the participating teachers often named was the way that the research fitted into research and society. The scientific relevance of the project was important because it helped the teachers connect the information and skills taught in school to the scientific world, and it worked as a great motivator. However, the participating teachers wished for a clearer connection to the research for which their students were gathering data. Most of the participants were only familiar with the topic through the news or social media. These publications were not directly related to the project, and the participating teachers stated that "it could be nice to be sent [new updates and publications] directly." This does not have to be an in-depth explanation of the research, but more on the level of showing what their data is being used for and receiving feedback from those involved in the research.

Societal relevance is important for teachers as well, as some of the participating teachers valued citizen science as a method to teach citizenship to their students. In the project this was seen in, for example, the modular NLT unit. An activity was included wherein students would take on different societal roles and go into discussion from that point of view. The teachers thought this was valuable, as they believe that they have "a task regarding [the improvement of] citizenship, [and] citizen science is, [they] believe, an important aspect of that in the sciences." This is supported by Maass et al. (2022), who discuss that modelling real-life problems could be used to connect science and citizenship education. Citizen science is one such way that could allow students to look at societal problems from new points of view.

In addition, the participating teachers valued that the data gathering of their students was being noticed by people in research and society, e.g. people involved in research and policymaking around the topic, for the purpose of motivating their students. The participating teachers wanted to be provided with opportunities for reward and recognition from outside the school environment. In the Snuffelfiets project this happened through presentation days where researchers and people involved in local government were present. Students and teachers were involved in the debate surrounding particulate matter and were asked to help formulate solutions for the posed problems. According to the participating teachers, this event had a great positive influence on the motivation of their students and themselves. The appearance of professionals involved in the topic worked very well, as one teacher explained that "no matter what kind of nonsense [the students] tell, I believe that a researcher earnestly listening to them has great added value."

8.5.4 Organisational Needs

Aside from the training needs as described by Lorke et al. (2019), there was also another need experienced by the participants that should be taken into account. Nearly all participants have said that participation in this project has taken extra time out of their already busy schedules in one way or another. In addition, the subject of particulate matter or research in general is not always within their areas of expertise, which made the project quite demanding. Nonetheless, all teachers seemed happy to give up their time for this project, as they were motivated by the project itself and the positive influence it had on their students.

In addition to the preparation time needed for the project, the participants also experienced a difference in the time they could allot to the project in general. One teacher noted that, given the different amount of time spent on the project and the different contexts in which it was executed, the project took on different forms as well. This particular teacher worked together with two students and only had 12 hours in total with the both of them to perform the research, which is a smaller scale than most other teachers were able to accomplish. Because they had less time and fewer participants, the amount of gathered data was less than they would have liked to have gathered. They expressed the need for ways to make the most out of as little time and participants as possible.

There is also the matter of personal prioritisation. One teacher said they did not have much time as a teacher in training as they were busy with their own studies, which made it difficult for them to take up many other activities. It was also said that, when a teacher changes schools, they first want to take the time to settle themselves before taking up large projects like a citizen science project. The turnover rate caused less enthusiasm within the school because established relationships between teachers were broken and new ones needed to be built.

As a solution to some of the problems regarding time and teacher turnover, the participants noted that they were missing some form of coordinator at their school. According to the participants, a coordinator could take charge of these kinds of projects and make the time for them to make sure they come to full fruition. They could keep the projects alive inside the school and help out the teachers who have less time on their hands. Another answer would be training individual teachers or groups of teachers to become these coordinators, so it could increase the efficacy of citizen science projects at schools.

Lorke et al. (2019) do not include the need for organisation in their list, but Harlin et al. (2018) include official support in their design claims. These claims, however, do not exactly line up, as the participants in this research were less concerned with the image of citizen science in the school, which the design claims seem to focus on, but more with the actual coordination of citizen science projects within the school.

8.6 Limitations and Further Research

It should be taken into account that all participants in this study have participated in the same citizen science project. As only participants in one project have been interviewed, only the needs specific to this project can be explored. Therefore, the full range of needs experienced by teachers across different kinds of citizen science projects remains unclear, e.g. the Snuffelfiets project is, at its core, a passive data-collecting project facilitated by the sensors, which is different from more active or participatory projects (e.g. where teachers and students collaborate with researchers in asking research questions). However, as seven teachers with a wide range of backgrounds and implementation methods have been interviewed, overall advice can still be given and general conclusions can still be drawn.

To gain a broader view of all the needs experienced by teachers while implementing citizen science, further research needs to be done on other kinds of citizen science projects. This is not limited to projects wherein students have a different kind of role, but also projects wherein teachers have a different kind of role or a different level of involvement as facilitator should be included. It should also be researched how exactly the suggestions of the participants could be implemented and whether they are effective. While the expert opinion of teachers holds sway, it would still be valuable to research the outcomes. For example, different voluntary recruitment approaches could be tested to evaluate what approach brings in the most participants in a formal education environment. In addition, it would be prudent to investigate whether the same needs can be found for citizen science projects outside the biological and environmental research domains.

8.7 Conclusions

During the interviews, it has been identified in what ways the participants have been supported in implementing an environmental research citizen science project like the Snuffelfiets project in their teaching, what they thought of this support, and how they would like to be supported in the future.

Firstly, teachers have received support for their role as facilitator of the project. While they already have experience as facilitators for their students, they would still welcome more support specific to the facilitation of the project, as depending on what they tried to accomplish with the gathered data, more support was needed to process the data. Secondly, the participants appreciated that the project and additional material were tailored to fit the level of the students, but they would like to have access to all available material and to receive more support in recruitment and engagement of their audience. Thirdly, the participants valued the connection to research and society through the presentation days organised by GlobeNL, but they still would like to receive more direct feedback and recognition from actors outside of schools to motivate their students and make them enthusiastic for science in society.

In conclusion, to answer the final research question, additions have been made to the proposed training needs framework for facilitators of Lorke et al. (2019) (Table 8.2). In general, the teachers often had the same kind of needs as described in the framework currently, so the framework still stands, but it could be expanded upon for teachers specifically. In this revised framework, the needs experienced during the Snuffelfiets project have been generalised, added, and categorised along the three proposed categories of Lorke et al. (2019). The category of organisational needs has been added, as teachers have named this often enough to warrant a new category, or at least a new need.

Appendix A: Interview Scheme

Interview Scheme Interviews Research Project

Translated from the original Dutch interview scheme.

Time: approximately 1 hour

Introduction *(5 min)*

1. To start slowly, briefly introduce myself and the project again.
2. Before starting the recording, reiterate that the (recorded) interviews will only be used for (this) research. Ask explicit permission for recording.
3. Start recording. State consent explicitly at the start of recording.

E.g. 'Thank you for your/your permission for this recording, ...'

Start *(5 min)*

Open the interview with easy questions to break the ice.
Could you introduce yourself?

- What subject(s) do you teach?
- How many years of teaching experience do you have?

How did you come into contact with citizen science?

- Why did you participate in this project with your class? What was your goal?

Topic 1: Core needs *(15 min)*
- In what ways was the science behind the citizen science project made clear?
- Was this information relevant to you and your class?
- Was there enough information to build your lessons upon?
- What about the content and/or form of the given scientific background would you like to change?

How was the relevance of the project made clear?

- Was the project and the work for the project made relevant for the students?
- Was a connection made to the curriculum? If so, in what way?
- Was the pedagogical relevance made clear? If so, how?
- Was this sufficient, or would you have liked more clarity on this?

Topic 2: Operational needs *(15 min)*
- How was the set-up of the project made clear?
- Was this relevant information?
- Did you add or remove certain things for your own class?
- What could have helped you or your students to understand this better?

How was the protocol for implementing the project made clear?

- Was it difficult to convey to your class?
- Were the students able to handle the instruments?
- Were adjustments needed for clarification?
- What could have helped you or your students to understand this better?

In what other ways were you supported in the implementation?

- For example, were lesson plans already in place?
- Were training sessions or preparatory lessons given to teachers?
- How did you experience that support? Was it useful for the learning experience of the students?

In what other ways would you have liked to be supported?

- Additional guidance (textual, communication, training)?
- Additional materials, instruments etc.?

Topic 3: Engagement Needs *(20 min)*
- How did you involve/recruit the students?
- Compulsory subject, voluntary participation, etc.?

- How did getting them involved go?
- Would you have liked more help with that, and how?

How did the students experience the project?

- Did the project match their views and interests?
- Was there room (facilitated by the project) for them to ask their own questions? And was that necessary?
- Did the level of the project meet their expectations or was it too much or too little?
- Were you supported in facilitating a good experience for your students? In what way?
- Do you think this support is necessary, or would you fall back on your own experience?
- What could have helped you to optimise the learning experience?

What did you think of the communication between the students and yourself and the communication between students during the project?

- In what ways was communication facilitated during the project?
- Did learners also need to communicate 'outwardly'? (For example, online or in the community).
- Were strategies for a communication approach given by the project's facilitators, or did you mainly fall back on your own experience?
- What could have helped you to make this communication run smoothly?

How did you experience the communication between yourself and the scientists and/or project group leaders?

- In what ways was there communication?
- Were you motivated by the project group?
- How quickly were you supported when you had questions?
- Would you like there to be more, less, or a different form of communication?

How did you acknowledge and reward the contributions of the students?

- Was this supported by the citizen science project? Or did you fall back on your own qualities and experiences?
- Had the project acknowledged, and perhaps even rewarded, the efforts of the class?
- Were your efforts acknowledged and rewarded?
- Did the pupils feel that their efforts were acknowledged and rewarded?

Were you approached for further involvement in citizen science projects?

- For example, learning more about citizen science in general, participating in other projects or the community around citizen science, or designing other projects.
- Would you like to be more involved in citizen science projects?

Closing

1. Thank the interviewee for their time.
2. Ask if the interviewee would like to see the thesis once it is completed.
3. Ask if the interviewee is familiar with other teachers who have done citizen science with students and are open to being interviewed.

Ask if the interviewee is open to further questions if clarification is needed.

References

Aono, Y., & Kazui, K. (2008). Phenological data series of cherry tree flowering in Kyoto, Japan, and its application to reconstruction of springtime temperatures since the 9th century. *International Journal of Climatology, 28*(7), 905–914. https://doi.org/10.1002/joc.1594

Aristeidou, M., & Herodotou, C. (2020). Online citizen science: A systematic review of effects on learning and scientific literacy. *Citizen Science, 5*(1), 11. https://doi.org/10.5334/cstp.224

Aristeidou, M., Lorke, J., & Ismail, N. (2022). Citizen science: Schoolteachers' motivation, experiences, and recommendations. *International Journal of Science and Mathematics Education.* https://doi.org/10.1007/s10763-022-10340-z

Bonney, R., Cooper, C. B., Dickinson, J., Kelling, S., Phillips, T., Rosenberg, K. V., & Shirk, J. (2009). Citizen science: A developing tool for expanding science knowledge and scientific literacy. *Bioscience, 59*(11), 977–984. https://doi.org/10.1525/bio.2009.59.11.9

Bonney, R., Shirk, J. L., Phillips, T. B., Wiggins, A., Ballard, H. L., Miller-Rushing, A. J., & Parrish, J. K. (2014). Next steps for citizen science. *Science, 343*(6178), 1436–1437. https://doi.org/10.1126/science.1251554

Denscombe, M. (2010). *The good research guide: For small scale social research projects* (4th ed.). Open University Press.

Finkel, A. (2018, February 7). How a German migrant planted citizen science in Australia – And why it worked. *The Conversation.* https://theconversation.com/how-a-german-migrant-planted-citizen-science-in-australia-and-why-it-worked-91385

GlobeNL. (2022, December 14). *Kans voor nieuwe scholen om mee te doen met de snuffelfiets.* Retrieved May 10, 2023, from https://globenederland.nl/onderzoeksprojecten/fijnstof/kans-voor-nieuwe-scholen-provincie-utrecht-snuffelfiets/

GlobeNL. (2023, April 18). *Fijnstof – GlobeNL.* Retrieved May 10, 2023, from https://globenederland.nl/leerlingen/onderzoeksprojecten/fijnstof/

Gray, S. A., Nicosia, K., & Jordan, R. C. (2012). Lessons learned from citizen science in the classroom. A response to "The future of citizen science". *Democracy and Education, 20*(1), 2. https://democracyeducationjournal.org/home/vol20/iss2/14/

Gura, T. (2013). Citizen science: Amateur experts. *Nature, 496*(7444), 259–261. https://doi.org/10.1038/nj7444-259a

Haklay, M., Dörler, D., Heigl, F., Manzoni, M., Hecker, S., & Vohland, K. (2021). What is citizen science? The challenges of definition. In K. Vohland, A. Land-Zandstra, L. Ceccaroni, R. Lemmens, J. Perelló, M. Ponti, R. Samson, & K. Wagenknecht (Eds.), *The science of citizen science* (pp. 13–33). Springer. https://doi.org/10.1007/978-3-030-58278-4_2

Harlin, J., Kloetzer, L., Patton, D., Leonhard, C., & Leysin American School High School Students. (2018). Turning students into citizen scientists. In S. Hecker, M. Haklay, A. Bowser, Z. Makuch, J. Vogel, & A. Bonn (Eds.), *Citizen science: Innovation in Open Science, society and policy* (pp. 410–428). UCL Press. http://www.jstor.org/stable/j.ctv550cf2.35

Het Grootste Onderzoeksprogramma Voor Scholieren – GlobeNL. (2023, April 5). Retrieved May 10, 2023, from https://globenederland.nl/

Jenkins, L. L. (2011). Using citizen science beyond teaching science content: A strategy for making science relevant to students' lives. *Cultural Studies of Science Education, 6*(2), 501–508. https://doi.org/10.1007/s11422-010-9304-4

Kelemen-Finan, J., & Dedova, I. (2014). Vermittlung von Artenkenntnis im Schulunterricht. Ergebnisse einer Befragung von Lehrpersonal in Österreich und bildungspolitische Relevanz. *Naturschutz und Landschaftsplanung, 46*(7), 219–225.

Kerson, R. (1989). Lab for the environment. *Technology Review, 92*(1), 11–12.

Kullenberg, C., & Kasperowski, D. (2016). What is citizen science? – A scientometric meta-analysis. *PLoS One, 11*(1), e0147152. https://doi.org/10.1371/journal.pone.0147152

Lorke, J., Golumbic, Y. N., Ramjan, C., & Atias, O. (2019). *Training needs and recommendations for citizen science participants, facilitators and designers* (COST Action 15212 Report).

Maass, K., Sorge, S., Ariza, M. R., Hesse, A., & Straser, O. (2022). Promoting active citizenship in mathematics and science teaching. *International Journal of Science and Mathematics Education, 20*(4), 727–746. https://doi.org/10.1007/s10763-021-10182-1

NLT. (n.d.). Retrieved May 10, 2023, from https://www.verenigingnlt.nl/

Pelacho, M., Ruiz, G., Sanz, F., Tarancón, A., & Clemente-Gallardo, J. (2021). Analysis of the evolution and collaboration networks of citizen science scientific publications. *Scientometrics, 126*, 225–257. https://doi.org/10.1007/s11192-020-03724-x

Pykett, J., Chrisinger, B., Kyriakou, K., Osborne, T., Resch, B., Stathi, A., Toth, E., & Whittaker, A. C. (2020). Developing a citizen social science approach to understand urban stress and promote wellbeing in urban communities. *Palgrave Communications, 6*(1), 85. https://doi.org/10.1057/s41599-020-0460-1

RIVM. (n.d.). *Burgerwetenschap | RIVM*. Retrieved 6 July 2022, from https://www.rivm.nl/burgerwetenschap

Robinson, L. D., Cawthray, J. L., West, S. E., Bonn, A., & Ansine, J. (2018). Ten principles of citizen science. In S. Hecker, M. Haklay, A. Bowser, Z. Makuch, J. Vogel, & A. Bonn (Eds.), *Citizen science: Innovation in Open Science, society and policy* (pp. 27–40). UCL Press. http://www.jstor.org/stable/j.ctv550cf2.9

Roche, J. A., Bell, L. C., Galvão, C., Golumbic, Y. N., Kloetzer, L., Knoben, N., Laakso, M., Lorke, J., Mannion, G., Massetti, L., Mauchline, A. L., Pata, K., Ruck, A., Taraba, P., & Winter, S. (2020). Citizen science, education, and learning: Challenges and opportunities. *Frontiers in Sociology, 5*, 613814. https://doi.org/10.3389/fsoc.2020.613814

Schmeller, D. S., Henry, P., Julliard, R., Gruber, B., Clobert, J., Dziock, F., Lengyel, S., Nowicki, P., Déri, E., Budrys, E., Kull, T., Tali, K., Bauch, B., Settele, J., Van Swaay, C., Kobler, A., Babij, V., Papastergiadou, E., & Henle, K. (2009). Advantages of volunteer-based biodiversity monitoring in Europe. *Conservation Biology, 23*(2), 307–316. https://doi.org/10.1111/j.1523-1739.2008.01125.x

Snik, F., Rietjens, J. H. H., Apituley, A., Volten, H., Mijling, B., Di Noia, A., Heikamp, S., Heinsbroek, R. C., Hasekamp, O. P., Smit, J. M., Vonk, J., Stam, D. M., Van Harten, G., De Boer, J., & Keller, C. U. (2014). Mapping atmospheric aerosols with a citizen science network of smartphone spectropolarimeters. *Geophysical Research Letters, 41*(20), 7351–7358. https://doi.org/10.1002/2014gl061462

SODAQ. (n.d.). *KnowYourAir*. Retrieved June 16, 2023, from https://knowyourair.net/

Tauginienė, L., Butkevičienė, E., Vohland, K., Heinisch, B., Daskolia, M., Suškevičs, M., Portela, M., Balázs, B., & Prūse, B. (2020). Citizen science in the social sciences and humanities: The power of interdisciplinarity. *Palgrave Communications, 6*(1), 89. https://doi.org/10.1057/s41599-020-0471-y

UC Davis Center for Community and Citizen Science. (2018). *Tracking sage grouse: In the field and online.* Retrieved June 29, 2023, from https://education.ucdavis.edu/sites/main/fles/ccs_yccs_case_study_sage_grouse_final.pdf

Part III
Fostering Engagement and Identity Through Citizen and Community Science

Chapter 9
Incorporating Community and Citizen Science into Elementary Schools: Key Program Design Features from Research and Practice

Jadda M. Miller, Heidi L. Ballard, Alexandra I. Race, Shulong Yan, Sol Henson, Evan F. Portier, and Emma R. Schectman

9.1 Introduction

Engaging young people in authentic scientific research through community and citizen science (CCS) initiatives presents rich opportunities for environmental education, but also complex design challenges. Effective CCS programs require navigating tensions between conducting meaningful research and providing impactful educational experiences grounded in locally relevant issues. In this chapter, we use the Our Forests program, focusing on 3 Key Design Features, as a model for developing a Youth-focused Community and Citizen Science (YCCS) program that links place-based research with science learning for elementary students. Our Forests serves as an example of how we designed for environmental science data collection and environmental science learning for elementary students.

Our Forests is a YCCS program that involves 3rd–5th grade students in forest ecology research and integrated curricular priorities in biology and environmental science education through its citizen science approach. Citizen science and related participatory approaches have become an increasingly common pedagogical strategy to provide authentic experiences for students to engage in scientific practices and connect their learning to local environmental issues (Ballard et al., 2017).

J. M. Miller (✉) · H. L. Ballard · A. I. Race · S. Yan · E. R. Schectman
University of California, Davis, CA, USA
e-mail: jaymiller@ucdavis.edu; hballard@ucdavis.edu; airace@ucdavis.edu; shlyan@ucdavis.edu; eschectman@ucdavis.edu

S. Henson · E. F. Portier
Sierra Streams Institute, Nevada City, CA, USA
e-mail: sol@sierrastreamsinstitute.org; evan@sierrastreamsinstitute.org

D. Tsybulsky, T. Tal (eds.), *Citizen Science Contributions to Biology and Environmental Education*, Contributions from Biology Education Research,
https://doi.org/10.1007/978-3-031-99298-8_9

Although "community science" is historically considered to be community-led research and monitoring, and "citizen science" is scientist-led, where members of the public primarily contribute through data collection (Shirk et al., 2012; Bonney et al., 2014), there continues to be debate in the U.S. and internationally about terminology in this field (Eitzel, 2021; Cooper et al., 2021). We refer to the overall range of participatory approaches as "community and citizen science" (CCS) to acknowledge and honor that they are distinct and separate participatory approaches to scientific research. CCS encompasses and accounts for the range of possibilities and interactions that occur between community members and professional scientists (Ballard et al., 2017).

The Our Forests program used CCS as a method for students to investigate their local forests while contributing forest data to collaborating forest managers (community partners). It was created and examined as part of a 5-year Research-Practice Partnership (Bevan & Penuel, 2017; Santagata et al., 2019) in which all aspects of both the program, curriculum design, and the educational research, were collaboratively developed by a community-based watershed organization (Sierra Streams Institute) and university-based education researchers (University of California, Davis), in close consultation with a County Office of Education and participating teachers, we were able to study and revise the program over 3 years of implementation in annual design cycles. This chapter includes design considerations and illustrative examples from our study of the program.

We draw on place-based and sociocultural learning theories as frameworks for grounding our curriculum. Place-based education foregrounds learning embedded within local contexts, fostering students' connections with their surrounding communities and environments (Smith & Sobel, 2010). This engenders opportunities for authentic, experiential learning and the potential for students to develop a sense of place (Kudryavtsev et al., 2012). Prior studies also indicate that place-based approaches to environmental science education can increase student motivation for acting on environmental issues by situating learning within students' lived experiences and local environments (Semken & Brandt, 2010). Sociocultural learning theory defines learning as a social process mediated by cultural contexts, with collaborative co-construction of knowledge through social interactions, tools, and shared activities (Vygotsky, 1978). In this context, scaffolding refers to providing support in the form of pedagogical techniques (e.g., materials, tools, and activity structures) for linking students' lived experiences with new information, allowing for greater engagement (Bouillion & Gomez, 2001). Additionally, when we refer to the idea of "co-construction of knowledge" we envision a joint creation of understanding through peer-to-peer dialogue and problem-solving, asking questions, and by collectively engaging in new hands-on learning experiences (Atwood et al., 2010; Damşa & Ludvigsen, 2016). This combination of place-based and sociocultural learning theories supports students' agency to act in personally consequential ways (Holland & Lachicotte, 2007). With this approach, we aim to better prepare students for engaging with environmental challenges.

9.2 Environmental Science Agency and the Data Life Cycle

Our learning goals for this program prioritized not only disciplinary science knowledge but also identity and agency with science as the key learning outcome (Ballard et al., 2017). This approach also supports young people in becoming agents of change (Freire, 2000) and draws on what we know about authentic participation in learning (Rogoff, 2003) and situated learning in practice (Lave & Wenger, 1991). Our focus on the 3 Key Design Features (Fig. 9.2) that framed and guided the Our Forests program, were drawn from previous research on ways to support science learning for students through YCCS, and specifically their development of Environmental Science Agency (ESA) (Ballard et al., 2017). ESA is defined as combining an understanding of environmental science content and inquiry practices, identifying with those practices, and enacting agency on behalf of the environment (Ballard et al., 2017). ESA has three core components: (1) Understanding environmental science content, (2) Identifying one's expertise in environmental science, and (3) Using that expertise as a foundation for environmental action. Ballard et al. (2017) examined the development of ESA in youth through two CCS programs for 11–18-year-olds that occurred in and out of school settings. This work identified key practices that supported youth in developing ESA, including taking responsibility for ensuring data quality, disseminating their scientific findings to external audiences, and investigating complex social-ecological systems. The Our Forests program intentionally built on these findings to structure the program to cultivate ESA in youth participants. We adapted these for elementary-aged students in school settings to develop the Our Forests 3 Key Design Features, where the program was designed to foster students' engagement in three practices repeatedly over the program: (1) Collecting place-based environmental data, (2) Analyzing and making meaning of their data, and (3) Sharing findings with community partners.

These 3 Key Design Features collectively make up the complete "data lifecycle" (Fig. 9.1., Bird et al., 2023), which Bird et al. (2023) suggest may build ESA when youth engage in not only data collection but also data analysis and sharing their findings with outside audiences. We emphasize this latter aspect when youth can express and practice their scientific knowledge by sharing their findings with, in the case of Our Forests, forest manager community partners, where youths' perceptions of their community partners become a critical component of their development of agency with environmental science data (ESA). Drawing from sociocultural (Brown et al., 1989; Lave & Wenger, 1991) and critical (Barton & Roth, 2004; Basu & Barton, 2009) perspectives, we argue the 3 Key Design Features play important roles in supporting youth to as Calabrese-Barton and Tan (2010) suggest, situate themselves within a scientific community of practice. By intentionally incorporating these research-based Design Features, the Our Forests program aimed to build ESA in youth participants through authentic scientific inquiry and experiences.

We are not suggesting that all 3 Design Features are necessary for learning gains. However, we suggest that they are all important. Many programs include only one or two of these design features and still achieve positive learning outcomes. We do

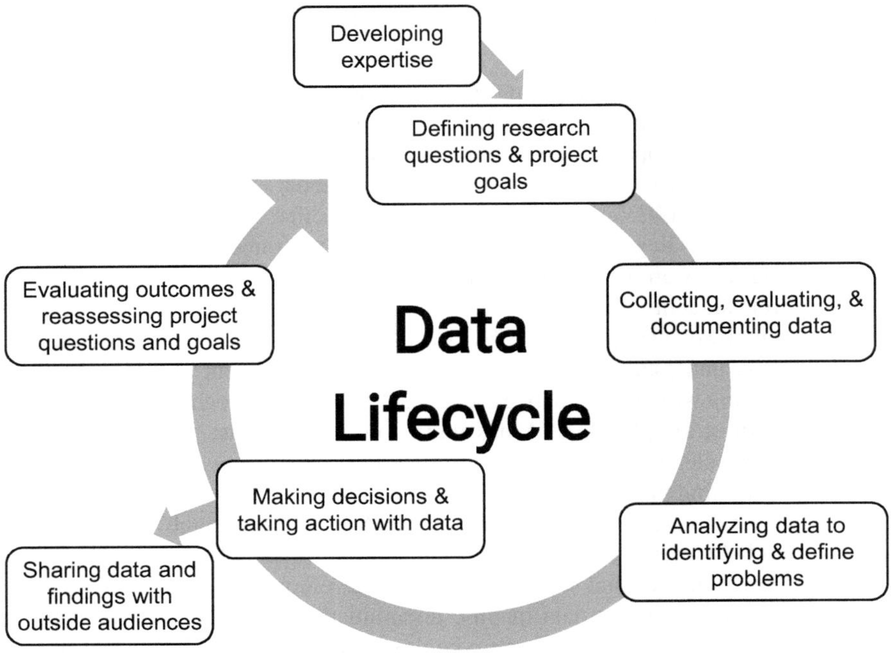

Fig. 9.1 Data lifecycle. (Adapted from Bird et al., 2023)

suggest that the integration of all three Design Features into CCS curricula and programming can create a powerful combination that enhances learning experiences and outcomes.

As part of the Our Forests scientific community of practice, our community partners have or will make forest management decisions around wildfire mitigation by incorporating the students' data. Using CCS for ecological monitoring is a common and growing aspect of natural resource management science (McKinley et al., 2017). As Bonney et al. (2014) argue, monitoring activities constitute a form of scientific inquiry, just as basic research does. By engaging with community partners in this way, we aim to highlight the scientific nature of the student's work and the role of CCS in informing natural resource management decisions. Of course, the student-collected data was not the sole or primary source of information for forest managers, but we here highlight how Our Forests created opportunities for students to contribute to ongoing scientific work in their community. By situating the project within the broader scientific community of practice, we seek to disrupt the binary distinctions between science learning and doing science, to empower students as active participants in the production of scientific knowledge (Harris et al., 2020).

9.3 Navigating Tensions in YCCS Design

At the core of designing effective YCCS initiatives for biology and environmental education in K-12 settings, is navigating the dual goals of providing impactful student learning experiences while also conducting authentic research or monitoring that is used beyond the classroom (Ballard, 2023; Kali et al., 2023). Careful attention must be paid to balancing these perceived competing priorities. We use the Our Forests program as an illustrative case study that provides research-based strategies and lessons learned for the design of effective YCCS programs in K-12 contexts.

As the world increasingly relies on the authority of science and scientific institutions to address environmental crises, we acknowledge that the culture and practice of science have been used to justify racism and to position particular ways of knowing, doing, and being as outside the realm of science (Liboiron, 2021), for adults and youth alike. In our project, we aimed to challenge these exclusionary practices by engaging students in authentic scientific work that is relevant to their local community.

Our focus on "authentic science" (i.e., what is useful to students and their community partners) is largely due to the implications it has for students' identity development (Harris et al., 2020). By participating in scientific practices that are meaningful to their community, students can begin to see themselves as capable of contributing to science and can develop a sense of agency in addressing environmental issues (Bang & Medin, 2010; Calabrese Barton & Tan, 2018). Therefore, throughout this chapter, we emphasize the words "authentic" which can be described as inquiry-based, project-based, or problem-based learning (Crawford, 2015) and "place-relevant" (connected to student's communities and local environment) to focus on the local importance of the science students are engaging with. Engaging in authentic and place-relevant scientific practices can deepen students' understanding and sense of connection to the scientific process (Harris et al., 2020).

9.4 Our Forests Program Description

The Our Forests program took place in the Sierra Nevada mountains of Northern California, USA as a co-led team effort consisting of University of California, Davis education researchers and Sierra Streams Institute, a community-based watershed organization. Our team partnered with the Nevada County Superintendent of Schools, 3rd through 5th-grade students, and their teachers, local forest scientists, and forest managers. This YCCS program focuses on designing forest ecology protocols around California's unprecedented increase in catastrophic wildfires in recent years due to historical fire suppression policies and the effects of climate change. Our Forests provided the opportunity for students to engage with this locally important issue. Collaboratively our team designed and developed forest ecology protocols for students to monitor and assess sites near their schools for fire risk.

Each year, approximately fifteen 3rd–5th grade teachers and their classes from 4 schools participated in the program (29 classes in total). The final iteration of the curriculum consisted of a year-long program that includes three all-day visits to the nearby Forest Investigation Zones (FIZ) specific to each school, a "share-out" lesson where students present to their community partners, and several classroom-based lessons for building foundational knowledge about the program and local forest ecology. Our 3rd through 5th-grade focus, addressed the need for more accessible elementary science teaching approaches that are aligned with Next Generation Science Standards (NGSS).

Over the 2 years of in-person program implementation, the UC Davis education researchers observed 18 of the participating classes as case study classes across the three grades at four schools. University researchers worked with each focal class teacher to select 5–8 focal students per class. Ethnographic observations of all field-based and in-class forest investigations were conducted as well as post-program semi-structured interviews (Merriam, 2014) with focal youth. Semi-structured interviews (30–60 min) provided student narration of their learning experiences (Creswell, 1996) to understand students' potential development of ESA. The examples of students' learning highlighted in this chapter are derived from these data sources.

9.5 Design Features

As mentioned above, Our Forests centers on 3 Key Design Features, (1) Collecting place-based environmental data, (2) Analyzing and making meaning of their data, and (3) Sharing findings with community partners. In the sections below, we explain the rationale behind each design feature, drawing on supporting literature. We then detail how each was implemented in the program, along with lessons and considerations for future program design.

Figure 9.2 describes each of the 3 Key Design Features, and how they ground students' learning in authentic science situated within a broader community of practice.

9.6 Place-Based Data Collection

YCCS experiences provide important opportunities for understanding science content through engagement in place (Ballard et al., 2017), such as when making field observations and collecting data. Many studies show evidence of science learning outcomes from first-hand experiences with data collection during CCS projects for adults, including increased understanding of specific ecological or science content (Brossard et al., 2005; Evans et al., 2005), science skills (Evans et al., 2005), or a commitment to carry out future stewardship activities (Crall et al., 2013). In CCS

Central Design Features	Explanations
1. Place-base Data Collection Youth collect data for projects related to local environmental issues.	Contextualizing science learning in place positions youth as doing "real" science that is meaningful and purposeful to their lives (Bouillion & Gomez, 2001; Calabrese-Barton and Tan, 2010; Calabrese-Barton 2012; Ballard et al., 2016).
2. Analyzing and Meaning-making Youth make sense of their own and others' environmental monitoring data.	Going beyond data collection to working with data supports students in the development of scientific practice skills and becoming critical consumers of data (Vahey et al., 2006; van't Hoof et al., 2012; Hug and Mcneill, 2008).
3. Sharing Findings Youth share their scientific results and recommendations for forest management and stewardship.	Sharing findings with community partners and forest ecologists supports students in building confidence and capacity for CCS work and provides an avenue for real-world impact (Bouillion & Gomez, 2001; Forbes & Zint, 2011; Pekar and Dolan, 2012; Monroe et al., 2016)

Fig. 9.2 3 Key Design Features

contexts, studies show how community-based wildfire prevention programs for youth (Ballard et al., 2012) use their local knowledge and lived experiences, including geographical, social, and cultural dimensions of place, to inform their scientific work (Bouillion & Gomez, 2001). This includes using their understanding of a place to contextualize their data findings (Taylor & Hall, 2013). This suggests ways in which place-based YCCS extends opportunities for youth to practice and apply their science content knowledge and skills.

In the Our Forests program, place-based data collection involved students going to nearby forest sites referred to in the program as Forest Investigation Zones (FIZ) to gather data on indicators of forest health. Place-based data collection took place across multiple site visits to their FIZ, focusing on a specific forest health protocol each time. Through these hands-on activities, students gained experience with scientific data collection protocols, including forest ecology measures like tree density, plant identification, and shrub coverage. The activities provided a real context for students to build connections to, and investigate, their local forests. The place-based component grounded the science learning in students' direct exploration of, and development of, a relationship with their local forests.

9.6.1 What Does Place-Based Data Collection Look Like

As described above, the place-based learning approach of the Our Forests program was designed to foster meaningful connections between students and their local environment in several specific ways. After instruction and practice, students in 3rd grade recorded data on mature trees throughout their plots by species and trunk size. Developing their strategies to identify specific tree species, we observed some 3rd graders gave names to these trees, like Bark and Bark Jr., highlighting the connections students began to make with nature and place. They also recorded the number and species of seedling and sapling trees and collected and identified tree seeds.

Place-based data collection for 4th graders involved measuring tree density using modified Biltmore sticks, identifying and marking different shrub species, using plant identification cards that also describe local Indigenous uses of the plants, and collecting data on understory plant diversity using a 1-m × 1-m quadrat to record the total number of tree, shrub, and understory plant species in their assigned plot. Several of these students conveyed their understanding of data accuracy when asked about the importance of correct data. For example, one student explained that they wanted to be careful because "inaccurate information could lead community partners to get the wrong information." This also demonstrated students' sense of responsibility and accountability to collect quality data that reflects taking on roles and identities with forest science.

Lastly, place-based data collection for 5th-grade students focused on more advanced forest ecology measures related to forest fuels that could burn in a wildfire. They collected data on tree diameters using diameter-at-breast-height tapes and dead and downed woody debris (DDWD) using a fuel sizing gauge, commonly referred to as a go-no-go gauge. They then collected data on shrub and canopy cover, gaining experience with forest estimation methods and tools including a densiometer and canopy scope (Brown et al., 2000), that help drive forestry management decisions. In post-program interviews, several 5th graders explained their connections between forest observations and their lived experiences. For example, a student noted, with concern, the similarity between the health of trees in their plot, which they saw as stressed or unhealthy, and the trees near their home. Making these linkages between what they learned during the CCS program and their home surroundings illustrates students situating the science content in the broader context of their lives.

9.6.2 Lesson and Design Considerations for Place-Based Data Collection

Summarized in Table 9.1 are specific design and pedagogical strategies that provide opportunities for students to develop a deeper identity with the science and connection to their place. In terms of program structure, we found that the timeframe (meaning the length of time spent in the forest and with the curriculum) is critical. Firstly, having multiple site visits allows students to become familiar with and

Table 9.1 Lesson and design considerations for elementary school CCS

To encourage student engagement in:	Place-based data collection	Data analysis and meaning-making	Sharing findings with community partners
Program Structure	Build in multiple site visits over an extended time allows for familiarity and attachment with the local environment Providing ample time during field investigations for students to make detailed observations and develop questions— this fosters curiosity Structured data-collection protocols per site visit	Dedicate time to intentionally building an understanding of what different numerical and visual representations mean Pair students across ability levels for peer learning and modify data analysis tasks for students who remain frustrated Making graphs by hand may increase students' ownership over data	Highlight how partners directly rely on student findings to make management decisions Build many touchpoints with community partners throughout the project After the end of the YCCS program share opportunities for students to join partners in implementing recommendations
Design of Activities	Structure protocols to scaffold complexity allows students to build skills slowly and become good at particular roles Providing real scientific tools, protocols, and instruction for using these tools with the opportunity to practice yields ownership in collecting accurate, quality data Balance structured data gathering and student-driven exploration Reflecting on learning experiences through journaling and drawing promotes deeper connections to content	Model analysis before students work independently builds their competence and confidence Reteach pre-requisite skills as needed alongside the data concepts Create activities that get students moving around to discuss differing perspectives Provide space for collaborative debate around what the numbers imply about the broader system Ensure students derive meaning from their findings through facilitated discussion	Demonstrate to students how their findings translate into forest conservation plans and the implementation Invite partners to join field investigations and assist with data collection During student presentations, encourage two-way dialogue versus one-way reporting Pose questions for collaborative problem-solving, putting community partner knowledge and student voices on equal footing
Pedagogical Approaches	Make local connections explicit—have students reflect on how the data links to prior knowledge and their community Embed repeated peer discussion and consensus-building into the data collection process Solicit and include community partners' real data needs to frame the purpose of data collection around addressing those defined problems increasing student motivation and comprehension of broader applications	Check frequently for understanding—not assuming students will intuitively know what representations mean Before data analysis, subtly assess students' baseline understanding through informal observation and conversations; tailor examples and explanations to the levels present in your group With guidance and intentional time spent making sense of data, students can gain higher-order skills for presenting evidence-based insights	It is important that students see community partners as an authentic audience they feel accountable to Establishing relationships with community partners early on allows partners to become familiar with students Continued participation pathways beyond the program, reinforce that students are valued members of a broader initiative to protect their local forests Prepare partners to receive student findings openly and respond to questions or misconceptions gently

connect to the local environment. Secondly, providing ample time during field investigations for students to develop questions, fostered their curiosity and allowed them to explore and discover subtleties at their site.

Additionally, we found that structuring protocols to scaffold complexity over time through guided activities, allowed students to build skills slowly and become good at particular roles. For example, students participated in exploratory activities where they first made observations and sketches, and then collected specific data on tree species. Similarly, providing scientific tools (e.g. binoculars, plant identification cards, etc.) and adequate instruction for using these tools, with ample opportunity to practice proper use, seemed to help students take ownership in collecting accurate data. Providing appropriate tools for investigation can empower students and build enthusiasm for data collection.

For in-the-moment pedagogical approaches, we found that making local connections explicit by having students reflect on how the data links to their prior knowledge and community can help solidify relevance. Embedding repeated peer discussion and consensus-building into the data collection process can help students construct deeper meaning and recognize the importance of comparing results.

Field-based data collection also poses challenges in the form of safety concerns for students in the forest, from poisonous plants to falling tree limbs. This can cause teachers and parents to be wary of students doing fieldwork. However, many of these concerns can be addressed by scouting sites where students will work, removing dangerous obstacles and plants, and communicating very clear boundaries and safety procedures to students before and during the program.

Above all, we suggest making sure to solicit and include community partners' real data needs to frame the purpose of data collection for students. Addressing these defined problems can increase student motivation and comprehension of broader applications. This makes the entire endeavor of data collection real, and as a result, students feel responsible for producing quality data.

9.7 Analysis and Meaning-Making

YCCS can create rich contexts for youth to work with the data they collect and further their science and data literacy skills (Butler & MacGregor, 2003). The NGSS Science and Engineering Practices at the elementary level require that youth have opportunities to work with data as they make and record their observations, argue with evidence, communicate their scientific findings, and apply scientific findings for problem-solving (National Research Council, 2012). These science and engineering practices can foster youth data literacy, which includes asking and answering data-based questions, interpreting information from data, evaluating data, and applying and communicating data to solve real problems (Vahey et al., 2006; van't Hooft et al., 2012). Hug and McNeill (2008) demonstrate that youth engage in science and engineering practices differently depending on whether they are provided with data or if they work with data, they collected themselves. This suggests different opportunities for increasing data literacy.

9.7.1 What Does Analysis and Meaning-Making Look Like

After collecting data on their local forests, 3rd-grade students participated in activities back in the classroom to analyze the data and derive meaning from it. For the tree density data, students counted the number of trees, reviewed totals as a class, and compared totals across different plots. For the seedling and sapling information, students made predictions about unmanaged forest areas based on their data, developed models of forest recruitment, and made bar graphs from their data. Students then made group consensus on estimates and entered findings into FieldScope (an online data collection and analysis platform).

Students in 4th grade engaged in collaborative and whole-class data analysis activities to identify patterns and derive meaning. They compared shrub diversity between managed and unmanaged forest plots and evidence-based claims about understory diversity in managed vs unmanaged areas. Throughout, they shared observations, deliberated discrepancies, and discussed similarities and differences between forest plots.

Fifth graders participated in collaborative analysis to identify patterns and develop evidence-based explanations. They reviewed tree diameter and DDWD data to present accurate information and compared their data between plots to check for consistency and settle on representative numbers. For shrub and canopy coverage, they calculated average values for the plot. Throughout, students articulated trends they noticed and crafted evidenced-based claims to describe their findings. As a capstone, they analyzed surface fuels, woody debris, and duff/litter data to generate explanations about fire behavior. These higher-order analytical activities required evaluating data across the broader investigation to construct meaning.

Students in all three grade levels reviewed their findings through data tables and visuals, checked for accuracy, discussed implications, and reached consensus. Comparing data between managed and unmanaged forest plots provided real-world context for drawing evidence-based conclusions and claims. Data analysis gave power and voice to students and helped them see that their work has a purpose beyond the project's end.

9.7.2 Lesson and Design Considerations for Analysis and Meaning-Making

Students were engaged in data analysis and meaning-making throughout the Our Forest program. We found collaborative data analysis activities helped students to derive meaning from their forest data and facilitated peer-to-peer learning– key takeaways are summarized in Table 9.1.

A particular challenge we encountered was the variation in student confidence in working with graphs and mathematics in a science context. As a result, we found that time should be dedicated to intentionally building an understanding of what these different numerical and visual representations mean. Additionally, pairing students across ability levels for peer learning and modifying data analysis tasks for students who remain frustrated can provide additional support if scaffolding attempts prove to be ineffective. Ultimately, the goal is for all students to feel capable of engaging with data in a way that is meaningful for them.

We found that modeling activities could play a key role in students making meaning of their data. Students were given models of forests and were asked to predict what that forest might look like in 20 years. Students were able to use their lived experiences, prior knowledge, and newly developed skills around collecting tree density and life cycle data to come up with their predictions. Students could then evaluate their predictions to determine if the future of the forest is going to be healthy or unhealthy, and at high or low risk for fire. This dynamic modeling activity allowed students to think through the implications of their data and what this means for the future of the forest.

Additionally, explicitly modeling analysis techniques step-by-step before students work independently, may help build their competence and confidence. Prepare to reteach pre-requisite skills as needed alongside the data concepts. Add small group pre-teaching for students needing more background knowledge before engaging in the data activities. We have also found that crunching numbers and staring at data tables can quickly lose students' attention. Bring data analysis to life through activities that get students up and moving around to discuss differing perspectives on findings and tackle discrepancies. Provide ample space for collaborative debate around what the numbers imply about the broader system. Rather than focusing on data collection alone, ensure students derive meaning from their findings through facilitated discussion, drawing conclusions, and tying patterns back to key concepts.

Lastly, show students that science isn't just about clear trends, unexpected results also have a story to tell. With guidance in analytical tools and intentional time spent making sense of data together, students can gain higher-order skills for parsing and presenting evidence-based insights.

9.8 Sharing Findings with Community Partners

Working with community members can benefit students, teachers, and community organizations. As many elementary teachers are uncertain about their ability to teach inquiry-based science (Forbes & Zint, 2010), working with local scientists can help overcome these barriers. Scientists can provide expertise and support to educators, while educators provide pedagogically appropriate transmission of knowledge to youth; this makes science content and skills more accessible to students (Peker & Dolan, 2012). Pitt and Schultz (2018) demonstrated how youth collected data with a YCCS program in partnership with National Parks and how National Parks used these data to inform environmental management decisions and actions. Community partnerships can provide youth with expertise and support. Additionally, Bouillion and Gomez (2001) argue that youth-based environmental monitoring projects should be mutually beneficial by collaborating on community-identified problems.

Throughout the Our Forests program, students were informed that the data they collected would be shared with community partners, (e.g., forest managers, fire scientists, and forest ecologists), to inform decision-making. After field investigations, educators prompted students to consider why their data matters. In the classroom, students entered their data into FieldScope, and instructors conveyed that partners are interested in their data to understand forest health. Before students' final presentations, they identified key points and recommendations to share with their community partners. Sharing their analyzed data, observations, and management suggestions gave purpose to students' work and built connections between their science learning and real-world applications. These activities aimed to develop students' understanding that science can be used as a tool for improving their community.

9.8.1 What Does Sharing Findings with Community Partners Look Like

Sharing data and recommendations with community partners reinforced real-world applications. In 3rd grade, students with a range of learning preferences actively participated in sharing forest data. This indicates that the share-out format provided an opportunity for every student to feel capable and that their opinions mattered.

In one 4th grade class, a debate over management ideas showed how preparing to present pushed students to synthesize their recommendations beyond the data analysis phase. Essentially, a few students advised controlled burns and tarping scotch broom (a highly invasive species).

Lastly, students in a 5th-grade class asked their community partner how they might continue their involvement in the Our Forests program beyond the school project. This community partner proposed working collaboratively with the students next year to integrate their suggestion to remove invasive plant species and plant more native species on their campus and in the local forest. The community partner valued and respected the students' forest management recommendations, illustrating the reciprocal nature possible through YCCS programs. Across grades, these examples demonstrate how sharing findings helped students to become empowered and see science as a tool for understanding and caring for their surroundings.

9.8.2 Lesson and Design Considerations for Sharing Findings with Community Partners

When structuring a CCS program, consider ways to demonstrate to students how partners may utilize their findings to inform management decisions. In the first iteration of the program, some students struggled to understand the context surrounding their data and how their community partners intended to utilize it. To address this issue in the second iteration, we provided more comprehensive framing at the beginning of the program and regularly reinforced this information throughout the program. By emphasizing the significance and practical applications of the data, we aimed to help students grasp the relevance of their work and its potential impact for the community partners. Fostering an understanding of this real-world application can further engage students. Additionally, collaboration is a core part of science, building multiple interactions in various ways with community partners throughout a project, not just a single visit, may prove beneficial. After program completion, there could also be opportunities offered for students to remain involved through optional monitoring, educational outreach, or other relevant science activities suitable for their age and interests. Ongoing partnerships like these can strengthen student connections to the scientific community and their sense of contributing meaningful insights. With thoughtful program structure and the creation of collaborative opportunities, students' work can become valued and integrated into broader environmental management efforts.

There may also be opportunities to show students how their findings could potentially inform forest conservation plans and on-the-ground implementation, where feasible. Demonstrating this real-world application can underscore the value of students' contributions. Also, consider inviting partners to join field investigations periodically and collaborate on data collection. This type of engagement may facilitate mentoring relationships and students may feel their knowledge gained from field investigations is valued. During student presentations, open-ended dialogue focused on students' interests and questions could supplement unidirectional reporting. Discussions structured around collaborative problem-solving may create space for community knowledge and student perspectives to interact on an equal plane. Activities that transparently connect student work to outcomes, while facilitating two-way partnerships, can strengthen the overall design of the CCS program.

When students know they are going to share findings with their community partners, they feel a greater sense of accountability if these relationships are established early on. Additionally, post-program, (optional) continued participation, paired with feasible implementation of recommendations, could reinforce students' perceptions that their role is meaningful and valued. Pedagogical foundations, such as recurring, collaborative interactions with community partners, may allow students to see science as an exchange within a larger community addressing local environmental challenges.

9.9 Conclusions and Contribution to the Field of Biology and Environmental Education

We have presented the Our Forests program as a case study for effective CCS programs in upper elementary school contexts. By detailing the program model, 3 Key Design Features, and research-practice-based recommendations, we aimed to provide considerations for both science learning and locally relevant research and monitoring with YCCS programs.

While beneficial in many ways, we identified several key considerations to potentially enhance each of the 3 Design Features and the implementation of YCCS programs broadly: (1) For place-based data collection, inviting local experts such as cultural practitioners and Indigenous knowledge holders to share traditional ecological perspectives could deepen students' understanding of and connection to the land, particularly if provided with historical context about the place; (2) Regarding data analysis and meaning-making: building in more scaffolded activities to explicitly teach different visual and statistical representations helps to strengthen data literacy. For example, allowing time for students to craft evidence-based explanations using claims and reasoning and then apply these findings to broader scientific concepts or issues deepens comprehension of what the data represents beyond numbers on a page; (3) For sharing findings with partners, training students in science communication formats could better equip them to effectively convey data, explanations, and recommendations. Having students craft take-home messages and tailor language based on the audience would also be beneficial. Lastly, closing the data lifecycle back to action could be integrated through structured opportunities for students to implement recommendations or conduct further research to answer lingering questions.

We encourage our research-practice-based strategies to be used by formal educators, outdoor educators, and scientists seeking to implement or enhance YCCS programs and environmental education. Our recommendations offer practical guidance for strengthening partnerships, scaffolding data literacy activities, and fostering ESA through an arc of investigation from collecting to applying findings.

When thoughtfully designed, YCCS programs offer tremendous potential for mutually beneficial collaboration across students, teachers, scientists, and community organizations. This chapter prompts reflection on the delicate balance between locally relevant research and impactful learning. When students are entrusted as valued contributors and see their data applied to local issues, they gain confidence in using science as a tool for understanding and improving their surroundings.

References

Atwood, S., Turnbull, W., & Carpendale, J. I. M. (2010). The construction of knowledge in classroom talk. *The Journal of the Learning Sciences, 19*(3), 358–402. http://www.jstor.org/stable/20799332

Ballard, H. L. (2023). Community and citizen science in schools: Nested inquiries, productive tensions, and the role of communities. *Instructional Science, 51*(5), 913–919.

Ballard, H. L., Evans, E., Sturtevant, V. E., & Jakes, P. (2012). The evolution of Smokey Bear: Environmental education about wildfire for youth. *The Journal of Environmental Education, 43*(4), 227–240.

Ballard, H. L., Dixon, C. G. H., & Harris, E. M. (2017). Youth-focused citizen science: Examining the role of environmental science learning and agency for conservation. *Biological Conservation, 208*, 65–75.

Bang, M., & Medin, D. (2010). Cultural processes in science education: Supporting the navigation of multiple epistemologies. *Science Education, 94*(6), 1008–1026. https://doi.org/10.1002/sce.20392

Barton, A. C., & Roth, W. M. (2004). *Rethinking scientific literacy*. Routledge.

Basu, S., & Barton, A. C. (2009). Critical physics agency: Further unraveling the intersections of subject matter knowledge, learning, and taking action. *Cultural Studies of Science Education, 4*(2), 387–392.

Bevan, B., & Penuel, W. R. (2017). *Connecting research and practice for educational improvement: Ethical and equitable approaches*. Routledge.

Bird, E. B., Ballard, H. L., & Harte, M. (2023). Data to decision-making: How elementary students use their community and citizen science project to reimagine their school campus. *Instructional Science*. https://doi.org/10.1007/s11251-022-09612-6

Bonney, R., et al. (2014). Next steps for citizen science. *Science (American Association for the Advancement of Science), 343*(6178), 1436–1437. Web.

Bouillion, L. M., & Gomez, L. M. (2001). Connecting school and community with science learning: Real world problems and school–community partnerships as contextual scaffolds. *Journal of Research in Science Teaching, 38*(8), 878–898.

Brossard, D., Lewenstein, B., & Bonney, R. (2005). Scientific knowledge and attitude change: The impact of a citizen science project. *International Journal of Science Education, 27*(9), 1099–1121.

Brown, J. S., Collins, A., & Duguid, P. (1989). Situated cognition and the culture of learning. *Educational Researcher, 18*(1), 32–42.

Brown, N., Jennings, S., Wheeler, P., & Nabe-Nielsen, J. (2000). An improved method for the rapid assessment of forest understory light environments. *Journal of Applied Ecology, 37*(6), 1044–1053. https://doi.org/10.1046/j.1365-2664.2000.00573.x

Butler, D. M., & MacGregor, I. D. (2003). Globe: Science and education. *Journal of Geoscience Education, 51*(1), 9–20. https://doi.org/10.5408/1089-9995-51.1.9

Calabrese Barton, A., & Tan, E. (2018). A longitudinal study of equity-oriented STEM-rich making among youth from historically marginalized communities. *American Educational Research Journal, 55*(4), 761–800. https://doi.org/10.3102/0002831218758668

Calabrese-Barton, A., & Tan, E. (2010). We be burning'! Agency, identity, and science learning. *The Journal of the Learning Sciences, 19*(2), 187–229.

Cooper, C. B., Hawn, C. L., Larson, L. R., Parrish, J. K., Bowser, G., Cavalier, D., & Wilson, S. (2021). Inclusion in citizen science: The conundrum of rebranding. *Science, 372*(6549), 1386–1388.

Crall, A. W., Jordan, R., Holfelder, K., Newman, G. J., Graham, J., & Waller, D. M. (2013). The impacts of an invasive species citizen science training program on participant attitudes, behavior, and science literacy. *Public Understanding of Science, 22*(6), 745–764.

Crawford, B. (2015). Authentic science. In R. Gunstone (Ed.), *Encyclopedia of science education* (pp. 113–115). Springer Netherlands. https://doi.org/10.1007/978-94-007-2150-0_144

Creswell, J. W. (1996). *Research design. Qualitative and quantitative approach*. Sage.

Damşa, C. I., & Ludvigsen, S. (2016). Learning through interaction and co-construction of knowledge objects in teacher education. *Learning, Culture and Social Interaction, 11*, 1–18. https://doi.org/10.1016/j.lcsi.2016.03.001

Eitzel, M. V. (2021). A modeler's manifesto: Synthesizing modeling best practices with social science frameworks to support critical approaches to data science. *Research Ideas and Outcomes, 7*, e71553.

Evans, C., Abrams, E., Reitsma, R., Roux, K., Salmonsen, L., & Marra, P. P. (2005). The Neighborhood Nestwatch Program: Participant outcomes of a citizen-science ecological research project. *Conservation Biology, 19*(3), 589–594.

Forbes, C., & Zint, M. (2010). Elementary teachers' beliefs about, perceived competencies for, and reported use of scientific inquiry to promote student learning about and for the environment. *Journal of Environmental Education, 42*(1), 30–42.

Freire, P. (2000). *Pedagogy of freedom: Ethics, democracy, and civic courage*. Rowman & Littlefield.

Harris, E. M., Dixon, C. G., Bird, E. B., & Ballard, H. L. (2020). For science and self: Youth interactions with data in community and citizen science. *Journal of the Learning Sciences, 29*(2), 224–263.

Holland, D., & Lachicotte, W. (2007). Vygotsky, Mead, and the new sociocultural studies of identity. In *The Cambridge companion to Vygotsky* (pp. 101–135). Cambridge University Press.
Hug, B., & Mcneill, K. (2008). Use of first-hand and second-hand data in science: Does data type influence classroom conversations? *International Journal of Science Education, 30*(13), 1725–1751. https://doi.org/10.1080/09500690701506945
Kali, Y., Sagy, O., Matuk, C., & Magnussen, R. (2023). School participation in citizen science (SPICES): Substantiating a field of research and practice. *Instructional Science, 51*(5), 687–694.
Kudryavtsev, A., Stedman, R. C., & Krasny, M. E. (2012). Sense of place in environmental education. *Environmental Education Research, 18*(2), 225–229.
Lave, J., & Wenger, E. (1991). *Situated learning: Legitimate peripheral participation*. Cambridge University Press.
Liboiron, M. (2021). *Pollution is colonialism*. Duke University Press.
McKinley, D. C., Miller-Rushing, A. J., Ballard, H. L., Bonney, R., Brown, H., Cook-Patton, S. C., Evans, D. M., French, R. A., Parrish, J. K., Phillips, T. B., Ryan, S. F., Shanley, L. A., Shirk, J. L., Stepenuck, K. F., Weltzin, J. F., Wiggins, A., Boyle, O. D., Briggs, R. D., Chapin, S. F., ... Soukup, M. A. (2017). Citizen science can improve conservation science, natural resource management, and environmental protection. *Biological Conservation, 208*, 15–28. https://doi.org/10.1016/j.biocon.2016.05.015
Merriam, S. B. (2014). *Qualitative research: A guide to design and implementation*. Wiley.
National Research Council. (2012). *A framework for K-12 science education: Practices, crosscutting concepts, and core ideas*. National Academies Press.
Peker, D., & Dolan, E. (2012). Helping students make meaning of authentic investigations: Findings from a student–teacher–scientist partnership. *Cultural Studies of Science Education, 7*(1), 223–244.
Pitt, A. N., & Schultz, C. A. (2018). Youth-based citizen science monitoring: Case studies from three national forests. *Journal of Forestry, 116*(2), 109–116. https://doi.org/10.1093/jofore/fvx008
Rogoff, B. (2003). *The cultural nature of human development*. Oxford University Press.
Santagata, R., Lee, J., & Sandoval, C. (2019). Research-practice partnerships in mathematics teacher education. In *International handbook of mathematics teacher education* (Vol. 3, pp. 183–210). Brill. https://doi.org/10.1163/9789004419230_008
Semken, S., & Brandt, E. (2010). Implications of sense of place and place-based education for ecological integrity and cultural sustainability in diverse places. In *Cultural studies and environmentalism: The confluence of ecojustice, place-based (science) education, and indigenous knowledge systems* (pp. 287–302). Springer.
Shirk, J. L., et al. (2012). Public participation in scientific research: A framework for deliberate design. *Ecology and Society, 17*(2).
Smith, G. A., & Sobel, D. (2010). *Place-and community-based education in schools*. Routledge.
Taylor, K. H., & Hall, R. (2013). Counter-mapping the neighborhood on bicycles: Mobilizing youth to reimagine the city. *Technology, Knowledge and Learning, 18*, 65–93.
Vahey, P., Yarnall, L., Patton, C., Zalles, D., & Swan, K. (2006). *Mathematizing middle school: Results from a cross-disciplinary study of data literacy.* Presented at the Annual Meeting of the American Educational Research Association, San Francisco, CA.
van't Hooft, M., Vahey, P., Swan, K., Kratcoski, A., Cook, D., Rafanan, K., Stanford, T., & Yarnall, L. (2012). A cross-curricular approach to the development of data literacy in the middle grades: The thinking with data project. *Middle Grades Research Journal, 7*(3), 19–33.
Vygotsky, L. S. (1978). *Mind in society: The development of higher psychological processes*. Harvard University Press.

Chapter 10
Flebocollect: School Citizen Science to Study Leishmaniosis Disease and the Impact of Ecosystem Disruption on Global Health

Rosa Gálvez, Ángela Bermejo, Marcos López de Felipe, Tarcisio Milagres, and Ana I. Mora-Urda

10.1 Citizen Science and Entomological Research

As a result of curiosity, novel questions are constantly raised by people around the world. Many of these questions have not already been answered or will never be resolved. Whichever the case, science is humans' response to these questions, for which experiments and hypotheses are employed to contrast our observations, a method inherited from European craftsmen dating back to the eleventh to seventeenth centuries (Clegg, 1979). The results obtained by these investigations may lead to new knowledge, which may open new lines of inquiry and more investigations and experiments. To better understand our study topic, a literature review is always needed prior, during, and after any research activity. Another requirement of any proper investigation is the acquisition of data, for which a lot of repetitions and time are required. Accordingly, much work may be needed to answer even what may appear to be the simplest of questions. Based solely on this concept, the idea of citizen science could be of great interest for researchers, especially those concerning organisms that are easily identifiable with a small amount of training, as is the

R. Gálvez (✉) · Á. Bermejo · A. I. Mora-Urda
Department of Specific Didactics, School of Education and Teacher Training, Universidad Autónoma de Madrid, Madrid, Spain
e-mail: rosa.galvez@uam.es; angela.bermejo@estudiante.uam.es; ana.mora@uam.es

M. L. de Felipe
Laboratory of Medical Entomology, National Centre for Microbiology, Instituto de Salud Carlos III, Madrid, Spain
e-mail: marcos.lopez@isciii.es

T. Milagres
Department of Microbiology, Immunology and Parasitology, Universidade Federal Do Rio Grande do Sul, Porto Alegre, Brazil

D. Tsybulsky, T. Tal (eds.), *Citizen Science Contributions to Biology and Environmental Education*, Contributions from Biology Education Research,
https://doi.org/10.1007/978-3-031-99298-8_10

case of insects (Kalaentzis et al., 2021). Data gathering across a large surveillance area may be easy for field citizen science researchers, as seen in previous research projects such as the European Butterfly Monitoring Scheme (EBMS). In this program, participants obtained essential data regarding Lepidoptera species around the continent (Swaay et al., 2008). Another example of insects and citizen science is the MosquitoAlert initiative, whereby citizens can upload pictures or other information concerning Culicidae mosquitoes (i.e. biting reports, breeding sites detections…) through an interactive application, being data later validated by taxonomy experts. This type of citizen science project is ideal for vector surveillance, and, in effect, this application played a key role in the first identification of *Aedes japonicus* (Eritja et al., 2019) and monitoring the dispersal of the Asian Tiger Mosquito, *Aedes albopictus* (Skuse, 1894), in Spain (Collantes et al., 2016). This invasive mosquito species has a great potential for disease transmission, as observed in several autochthonous dengue outbreaks in the country linked to this mosquito (Aranda et al., 2018; Monge et al., 2020). Another key aspect of citizen science in research projects is the accessibility of researchers to data from more private areas, which may be highly difficult to access during traditional field research, as was the case of the NESCOTIGER project (Novel Strategies for the Control of the Asian Tiger Mosquito), where control tools and surveillance traps were installed and collected by citizens in their residential areas and later evaluated by researchers (Gil et al., 2022; López de Felipe et al., 2022). Nevertheless, data gathering is not the sole role that citizens may play during scientific research interventions, as they may also be part of the design of the actual scientific projects, and even raise novel questions (Bonney et al., 2016; Hadjichambis et al., 2023). As such, citizen science needs to be seen as a symbiotic relationship, with benefits for both citizens and researchers alike. Such is the case of the Flebocollect citizen science project (de Freitas Milagres et al., 2025).

10.2 Flebocollect: The-Making-of

The citizen science group Flebocollect was created in response to the World Health Organization's (WHO) demands for sustainable and innovative solutions to reduce the impacts of vector-borne diseases like leishmaniosis (WHO & UNICEF, 2017). Phlebotomine sand flies are the sole vectors of *Leishmania* species, causing leishmaniosis in both animals and humans. As such, their study is essential to control these zoonotic diseases, which means those transmitted from animals to humans and vice versa. In this context, it should be emphasized that leishmaniosis is considered as emerging or re-emerging diseases. Nevertheless, they are also seen as neglected tropical diseases, or maladies tightly associated with poverty (Alvar et al., 2006), while not getting in general much attention by public services or researchers. Sand flies are crepuscular-nocturnal insects showing positive phototropism (they are attracted to light), so the use of CDC (Center for Disease Control and prevention) light traps is common practice in phlebotomine surveillance programmes (Alexander, 2000). Although effective, these traps are too expensive, making phlebotomine sand fly surveillance inaccessible to many communities, educational projects, or scientific research groups with limited budgets. Hence, the Flebocollect project arose from the idea of developing a DIY (Do It Yourself) light trap model to capture phlebotomine sand flies employing reused cheap materials as an alternative strategy to raise awareness about leishmaniosis and vector monitoring (Gálvez et al., 2022b). However, a long road has been travelled to reach that goal (Fig. 10.1).

Originally, the project was part of the Bringing Research Into The Classroom initiative (BRITEC), funded from 2018 to 2021 by an Erasmus+ KA2 Programme under Key action 2, Cooperation for Innovation and Exchange of Good Practices (KA2 n° 2018-1-PL01_KA201-051080). This project, involving cooperation between schools and research institutions, was designed to introduce citizen science activities into primary and secondary schools in the academic year 2020–2021. The aim of the project was to spark the interest of young students in science through their direct involvement in scientific procedures, interaction with scientists and the use of high-quality educational materials.

Spanish partners of the Universidad Autónoma de Madrid (UAM) designed in 2019 the subproject Flebocollect as part of the BRITEC initiative. The main goal was to involve children in citizen science activities in and around the classroom to study sand fly populations in the Madrid region (Spain) (Gálvez et al., 2019). This first Flebocollect team originated as an open collaborative working group seeking to develop a DIY light trap model. The multidisciplinary group met every 2–3 weeks from May to December 2019 in the Fablab (Fabrication Lab) of a publicly funded citizen laboratory in Madrid (Spain), Medialab Prado (MediaLab Prado, 2021).

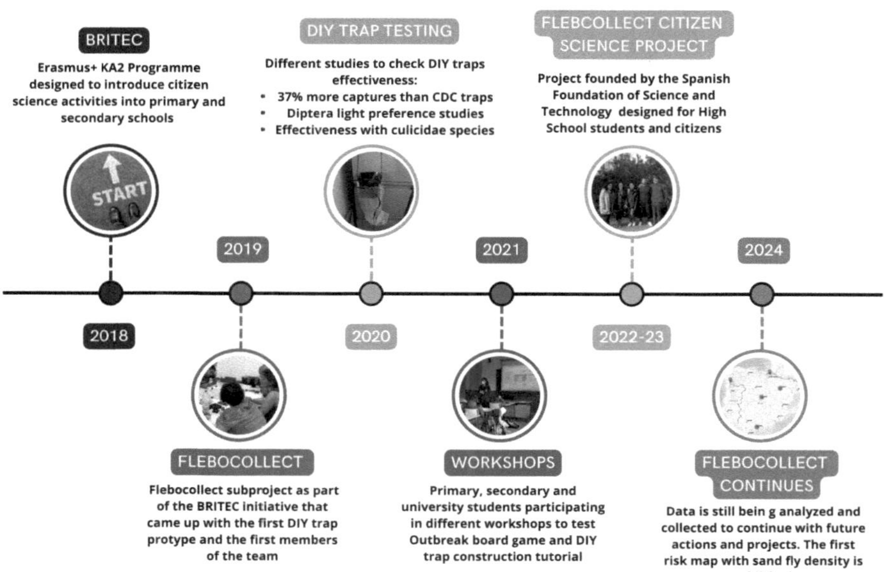

Fig. 10.1 Flebocollect project timeline. (Own creation)

Collaborators were 21 individuals; 6–7 people on average participating in each two and a half hour meeting held at different times on 13 occasions. The educational level of participants was diverse: biologists, fine arts graduates, physicians, veterinarians, university teachers of non-scientific disciplines, high-school teachers of biology and technology, biology and veterinary science students, etc. This multidisciplinarity within the group greatly enriched the approach to this project. Finally, the first prototype of DIY light trap was out and a step-by-step guide for making them was provided to anyone who was interested (Gálvez et al., 2022b). Flebocollect light traps consist of a small electronic circuit with a diode light that attracts phlebotomine sand flies into a suction tunnel made from reused materials, a milk/juice tetra brick, and a pc computer fan. Sand flies are directed into a mesh net (e.g., fruit bags, dirty clothes' bags). Finally, to hang the trap, an old shoelace is threaded through holes made on opposite sides of the suction tunnel (Fig. 10.2). Flebocollect traps are cheap and easy to manufacture, which makes them a great option for research and to use them in citizen science projects. Moreover, by reusing old materials, it is a good way to support circular economy.

Next, Flebocollect actions consisted of developing the final Flebocollect DIY trap, after the designing of several prototypes, and also testing it in a preliminary field studies. The first one, was carried out in the Community of Madrid (Central Spain) during the summer of 2020 (Gálvez et al., 2022b). During this research, Flebocollect traps were shown to obtain higher mean capture rates (37%) than these commercial models (CDC light traps), though, no statistical differences were observed among traps. Besides, investigators studied which light wavelength would yield the best capture rate was tested (López de Felipe et al., 2023).

In fact, since 2021, several DIY light trap workshops have been held for individuals from primary to secondary schools and postgraduate education (Gálvez et al., 2022a). Apart from being of interest as a research tool, especially for scientific research groups with limited budgets, DIY light traps were also intended for

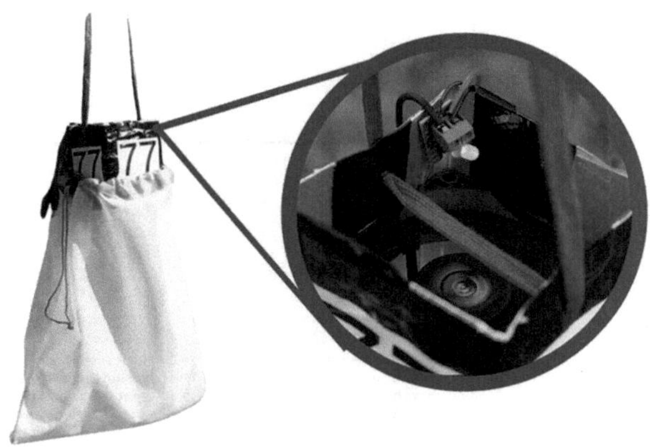

Fig. 10.2 DIY phlebotomine sand flies light trap consisting of an electronic circuit, suction tunnel supporting a fan, and collection bag

use in community science projects and education projects. However, once the final prototype of the traps was obtained and very positive results were obtained in terms of its effectiveness, potential and educational purpose the researchers requested a larger-scale citizen science project, which would be described in the next section of this chapter.

10.3 Flebocollect Citizen Science Project (2022–2023)

From July 2022 to September 2023, the Flebocollect citizen science project was financially supported by the Spanish Foundation of Science and Technology (reference number: FCT-21-16782) with the aim to implement educational activities mainly in high schools, but also for citizens, for the promotion of health education and contribution to scientific investigations. In addition, Flebocollect team grew up and new members started to collaborate. It is characterized by its multidisciplinary nature, so members have an extensive scientific training in diverse fields like biology, entomology (Gálvez et al., 2020), as well as in didactics of experimental sciences. Furthermore, most of the group members belong to the Universidad Autónoma de Madrid, being teachers, investigators and PhD students. What is more, other professionals in this field act like close collaborators like veterinarians, teachers and biologists from other institutions.

This Flebocollect project focuses on health education, specifically in the context of zoonotic diseases such as leishmaniosis. In countries with disadvantageous socioeconomic, these vector-borne diseases are a major cause of mortality and/or morbidity and cause substantial economic losses. Because of global changes (i.e., globalization, mass population movements, climate change...), many of these diseases are re-emerging or being newly introduced to previously disease-free areas. Examples of the re-emergence of old and introduction of new diseases can be found in the leishmaniosis outbreak of 2009 in southern Madrid (Arce et al., 2013) or as in the several autochthonous dengue cases reported in Spain in the last 10 years because of the introduction and establishment of the Asian Tiger mosquito, *Aedes albopictus* (Aranda et al., 2018). However, while education has been identified as one of the key tools for disease control, research on the effectiveness and impact of health education as a direct intervention is limited (Butala et al., 2021). Pérez (2020) stated that school interventions on health and disease issues must be integrated and contextualized in a set of actions or contents that are close to the students. This is why the approach of the Flebocollect education project was designed within the framework of the human leishmaniosis outbreak that occurred in a southern area of the Community of Madrid at the end of 2009. This outbreak was the largest recorded in Europe to date, as a consequence of a change in land use and abundance of these vectors that spread the disease, not only among wild animals such as rabbits and free-ranging animals, but also to humans (Vilas et al., 2012).

Having described the importance of the Madrid leishmaniosis outbreak, added to the lack of knowledge of the population of this place about this disease and the low awareness to control the vector and prevent new outbreaks, this new project was developed for students and citizens alike. The Flebocollect education project consisted on two modalities:

(I) Line 1, the complete project for educational centres in the Community of Madrid (all activities)
(II) Line 2, for education centres in other Spanish provinces and citizens (not all the activities).

The combination of results from both modalities allowed the creation of a risk map of sand flies' distribution and abundance all over Spain (Fig. 10.3; Fig. 10.5). According to citizen science taxonomy proposed by Bonney et al. (2016), this proposal does not fit into a specific category, it presents a hybrid character. From one point of view, this can be considered as a "Data collection project", as participants installed the traps and notified scientist of their captures, so they can carry out proper scientific research. Nevertheless, it can be considered a "Curriculum-based project", as most of the participants were students, it is supervised by their ordinary teachers, and it is aligned with state and national curriculum standards.

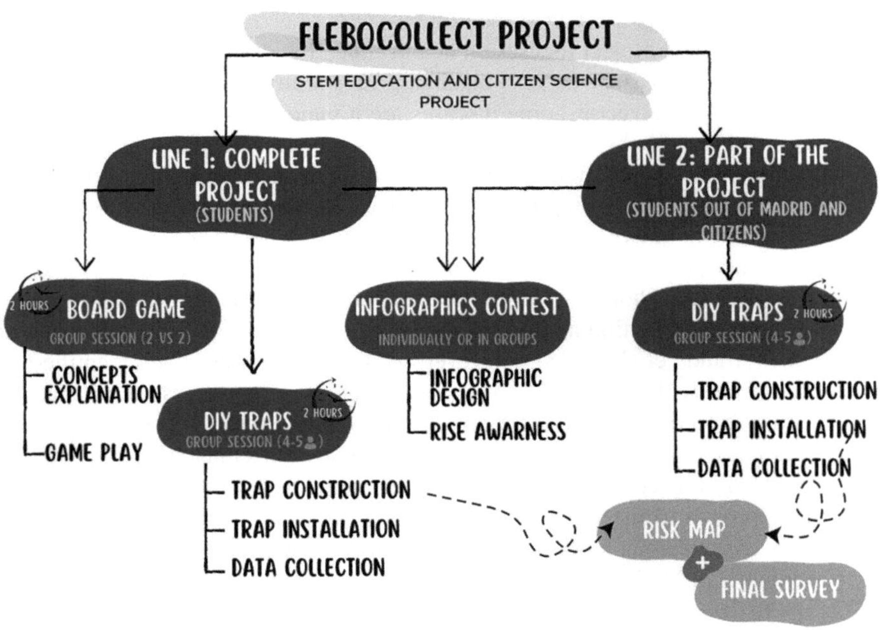

Fig. 10.3 Activities routes that make up the Flebocollect project divided into two lines or modalities: (**a**) Line 1 refers to the complete didactic project; (**b**) Line 2 refers only to part of the activities. Trap construction and installation allow the develop of a map of phlebotomine sand flies density to assess the risk of new outbreaks of leishmaniosis

10.3.1 Line 1: Complete Flebocollect Project

As it was mentioned before, this modality of the project was developed in educational centres of the Community of Madrid (Central Spain). The recruitment process was made through publicity in social media: Instagram, X (previously called Twitter) and Facebook.

All educational centres interested in participating had to fill out an online questionnaire where they had to indicate their educational centre name, the specific students with whom they wanted to participate (age and educational level) and the motivation they had to do so, as well as how the educational route would be developed in their centre (teachers and school subjects involved). The questionnaire also included some basic rules to be able to participate in the project. Participation was only allowed to high-school students (12–18 years old) accompanied by science teachers, or with certain training in this field, and located within the limits of the Community of Madrid. This questionnaire served as a filter to select the educational centres that showed greater involvement and motivation with the project. Likewise, an attempt was made to normalize the trap distribution, and educational centres were selected from different areas of Madrid and with students from different educational levels. Exceptionally, two Primary Education schools (5th and 6th grade of Primary, 10–12 years) were allowed to participate to find out if this educational project could be carried out, in a near future, with students of that level.

As this modality of the project was developed with students, some methodological considerations were taken into account, as this didactic sequence aimed not only to collect data from students captures (i.e. epidemiological interest of the project), but also to increase their scientific literacy and interest in entomology and science research. Game-based learning is one of the active methodologies that has yielded best results, as many education theories, such as constructivism (Phillips, 1995), situated learning (Lave & Wanger, 1991), and problem-based learning support their use (Savin-Baden & Major, 2004). In addition, game-based learning can enhance students' skills such as decision-making, problem-solving, and teamwork (Kolb & Kolb, 2005), all of which are mentioned in the Spanish education curriculum (RD 217/2022). Games have also proved to motivate students to learn more about each specific game's topic (Jääskä & Aaltonen, 2022). Furthermore, while one of the recurring topics of biology games is genetics (Da Silva Santos et al., 2022), there are some examples of games covering vector-borne diseases. Hence, the literature contains descriptions of board games on pest control (de la Vega et al., 2022), insects (Cosme et al., 2020; Khelifa & Mahdjoub, 2021), vector-borne diseases (Amelia & Setiawan, 2019), and even a game on leishmaniosis (Alidosti et al., 2022). It should be noted that the great majority of the examples of biology education board games are from countries in which Neglected Tropical Diseases (NTD) have a great impact (Butala et al., 2021). Moreover, authors also chose the STEM approach (from the acronym Science, Technology, Engineering and Mathematics). This methodology is a recent phenomenon in the Spanish educational context (Lupión-Cobos et al., 2023) and it is seen as a promise of innovation and preparation for the future (Couso, 2017). This methodological approach STEM education focuses on promoting active learning for students, where the approach to the different phenomena to be studied or problems to be treated, in addition to being carried out in an interdisciplinary manner, is carried out in the same way as it is presented in nature. Also, it has been described as an effective tool to generate curiosity and interest in students' everyday problems (Cilleruelo & Zubiaga, 2014). The Flebocollect project is classified as a STEM approach, as students had to obtain information from graphs, tables, or diagrams (mathematics), and must deal with different medical entomology and health concepts (science) and ways of trapping insects (technology-engineering).

The educational objectives of the Flebocollect project for students were to solve real life problems such as the Madrid leishmaniosis outbreak and to spend time in the shoes of entomologists specialized in the study of phlebotomine sand flies. This line 1, or complete project, consists in a didactic sequence composed of the following activities and timeframes for them (Fig. 10.3).

10.3.1.1 Playing the Board Game Called "Outbreak: The Strange Case of the Leishmaniosis Outbreak in the Community of Madrid" (Fig. 10.4)

Students were divided in groups of 4 people (2 teams of 2 children each) that competed to solve the mystery of the game during 2 hours of guided work. This board game was created by the Flebocollect team through the inclusion of real conducted research activities and scientific published papers concerning the leishmaniosis Madrid outbreak (Bermejo & Gálvez, 2023). It was inspired by the famous board

Fig. 10.4 Board game "Outbreak: the strange case of the leishmaniosis outbreak in the Community of Madrid"

game called Clue. Two students' groups must compete for being the first to complete all the questions of the "Research notebook" (vector, reservoir, type of leishmaniosis, age and gender of the infected population, outbreak starting year, etc.) thanks to different cards with graphs, tables, texts that they obtained by moving through the game board. During this activity, Flebocollect investigators were there to help students with some difficult concepts such as lab techniques or medical vocabulary and, also guiding them to interpret the cards. When the students finished the game, a group session was held in which it was expected that students would understand the consequences of anthropogenic alteration of the environment on public health.

10.3.1.2 Building DIY Light Traps to Capture Phlebotomine Sand Flies

Students were divided in groups of 4–5 people that followed an explanation given by the Flebocollect research team to build the DIY trap (Fig. 10.2) during 2 hours of guided work. A video tutorial of how to build the trap was projected in every class screen and all the materials were provided, except in the case of the tetra brick and the shoes laces, with the intention of making students more involved in the construction of the trap and promoting awareness of recycling and reusing materials. During this activity, Flebocollect investigators were there to help students in everything they needed, for example, to aid them weld the electrical components. At the end of the activity, students received a brief explanation of how to use the traps and an online document or guide for them to solve any question (Online materials).

10.3.1.3 Installation of DIY Traps and Recording Details of Sand Flies Captured Using the Flebocollect App

This part of the activity involved autonomous work by children and their supervising adults (teachers, parents…). Students started this phase of the project by installing the traps at the end of May, at school's surroundings, although they could also choose themselves the specific sampling sites: school's surroundings, gardens and terraces of their homes or other areas where they had second homes such as open countryside, large farms or small villages. Their ordinary teachers explained them how to use the app and registered the date collected with the traps (Flebocollect App, 2023), as they previously received an online formation by the Flebocollect team. Moreover, the app is easy to use. After registering the trap in this Flebocollect trapping App, the location was automatically uploaded. They had also to add a picture for investigators to show the epidemiological scenario of the trapping sites. Traps were installed by students from dask to dawn, after which they turned them off and gathered the traps, which were frozen in their freezer to kill the captured sand flies. Students also counted the captured individuals following indications of a guide designed for them, recording this data in for investigators to check their responses. Captures were also sent to the Flebocollect team by post-mail for taxonomical identifications.

10.3.1.4 Designing infographics to raise awareness of the Madrid outbreak of leishmaniosis and to participate in a contest

Autonomous work had to be done by the students. It was not a compulsory activity, so every student that wanted to explore his creativity, individually or in groups (up to 5 people per group), was welcomed to participate in the contest. To join, students were asked to prepare an infographic through a computed concerning sand fly. Prepared materials were sent to the Flebocollect group through e-mail, and we coded them to do an anonymous voting. There were 3 prices for the top 3 infographics, which consisted of a USB microscope and two board games about science.

10.3.1.5 Final Survey

A final survey was sent to every participant to analyze the following aspects:

1. Participants perception about citizen science
2. The project's impact on their science perception
3. Students' future action in citizen science projects
4. Personal assessment of the project

Results from this survey will be later explained (see Sect. 10.5). Participants had to complete an online questionary once they had completed all the activities. Questionaries were sent in May 2023 and responses were collected until the end of September 2023.

10.3.2 Line 2: Part of the Flebocollect Project

This line 2 is considered "part of the project" even though it excludes the first activity, the board game. Furthermore, participants are not students from the Community of Madrid, but rather pupils from educational centres outside this region. In addition, the general public all over Spain was also invited to take part in this line. The recruitment process was much simpler than for line 1, as anyone that contacted us through social networks or e-mail and did not belong to a Madrid education school was invited to participate in this line 2 modality until resources were exhausted. Also, the Natural Science Museum of Madrid made the announcement.

In the case of students out of the Madrid area (same ages and educational level as in modality 1), they received in their high schools all the necessary materials and tutorials for the construction of the traps, while their science teachers oversaw the developing of the traps construction workshop with the aid of the construction video tutorial made by the Flebocollect research team. Likewise, teachers attended an online training webinar to later better explain to their students the use and management of the traps as well as the Flebocollect App. Regarding the infographic design activity, it was carried out in the same way as in modality 1, since it does not require the presence of Flebocollect researchers. On the other hand, as for the public, several workshops were held at the Natural Science Museum of Madrid, so that those citizens who were interested in participating could build their traps, learn how to use them and thus also be able to contribute to the project with their catches that they would notify, also, through the app. Citizens only had to register for the activity through the museum's website

10.3.3 Final Results and Participation

At the time of writing, 20 schools have participated in the project, 14 from the Madrid area and the rest from other Spanish autonomous communities (Guadalajara, Granada, Alicante, Huesca, Zaragoza y Málaga), which means around 1800 students of many different ages (10–18 years). Additionally, over 350 Flebocollect traps have been constructed by students and citizens, and they have been capturing phlebotomine sand flies in their study areas during the 2023 summer season.

All participants, both from lines 1 and 2, through the mobile phone Flebocollect App, registered and identified their captures with the help of didactic materials including basic information to identify sand flies and a traps handling manual. Then, they sent us the captured insects for expert validation of the presence of sand flies and a posterior taxonomic identification to the species level. The scientific goal was to generate a risk map of phlebotomine sand fly distribution that would help monitoring the risk of leishmaniosis transmission in Spain. In this way, by engaging community members in citizen science, not only do researchers gain valuable data, but the public also gains scientific knowledge (Bonney et al., 2016). Data is still being analysed but a preliminary risk map is available to the public, showing captures the participating students have reported trough the app (Fig. 10.5). This map is constantly updated, as captures continue to be received and validated, and are periodically being included in the map, even though the financing of the project by the Spanish public administration has ended.

Fig. 10.5 Risk map showing the sand flies distribution in Spain. It can be consulted online in the link: https://www.flebocollect.com/mapa/

10.4 Engaging the Public in the Flebocollect Project

An important step of the Flebocollect citizen science project was its national dissemination, carried out through a series of activities designed to bring the project closer to the general public and to also contribute to the scientific field itself. The transmission of scientific data can play a crucial role in the success of a project. Communicating the objectives, results, and impacts of a project in a clear and engaging way, helps to popularize scientific projects, awakening the public's interest and curiosity in relation to science, and leads to their engagement, participation, and collaboration (Danilina, 2022). Moreover, effective knowledge dissemination is essential to share ideas and developments with other researchers and professionals in the field, helping to establish the relevance and importance of the project, in addition to facilitating the exchange of knowledge and expediting the development of innovative methodologies (Weingart et al., 2021; Danilina, 2022). Anyhow, being the project's core, the Flebocollect traps, a result of a citizen science initiative, and traps trapping sites depended soley on the participants' criteria, citizens are a major key part in the development and designed of the project itself.

To this end, a website was created containing essential information such as the functioning and purpose of the project, how to participate, and basic concepts about sand flies, STEM methodology and citizen science (Flebocollect Web, 2023). The website is user-friendly to guarantee a pleasant and intuitive user experience. Thus, the public can easily find what they are looking for and explore different aspects of the topic. Through the website, users can also follow updates and access step-by-step descriptions of the work being conducted, besides subscribing to a newsletter, with all this information available directly in their personal e-mail. It is also possible to access publications (book chapters, articles, and conference papers) generated by the Flebocollect group. Importantly, the project team can be contacted. The website works as a central platform so that details of the project stages can be shared allowing for a more complete presentation of the content we produce. It is a reliable database for interested individuals who want to learn more about Flebocollect.

Another national dissemination strategy was the creation of pages on social media such as X (Flebocollect X, n.d.) and Instagram (FlebocollectIG, n.d.). This type of platforms offers a broader and more global outreach, such that information can be transmitted to a larger audience, promoting the dissemination of knowledge beyond traditional borders (Huber et al., 2019). In addition, social networks helped us present the Flebocollect project in a more attractive way through visual resources such as videos, images and infographics, simplifying complex concepts and making them more understandable to a non-specialized audience. Another benefit is that the population can directly interact with the project through the hashtag (#flebocollect) and can share photos, videos and other types of online posts. Likewise, it is possible to interact with users by sharing their publications and indirect posts such as reminders of commemorative dates, bringing a feeling of closeness to the participants, and stimulating the engagement of interested individuals.

An important point for the popularization of the project are the workshops offered and the participation of the research team in science fairs events, such as sessions conducted in the Natural Science Museum of Madrid. Museums are generally considered places of prestige and trust, and their association with science and education can increase the credibility of the scientific projects presented (Liebert, 2019). These events provided an interactive and educational environment where the project could be presented to the public in a more direct and participatory way. Visitors had the opportunity to immediately get involved with the programme: building DIY traps, playing the board game "Outbreak", seeing prepared sand flies slides, among other activities. In addition, visitors of different age groups had the opportunity to ask questions and obtain detailed explanations in person. By viewing the "science in action", citizens can visualize and experience theoretical concepts in a practical context. This creates a stimulating learning environment, sparking visitors' interest and curiosity. To also emotionally connect with the public, during the workshops, children were given personalized t-shirts (Fig. 10.6) with the project logo and "Junior Entomologist" name tags (Fig. 10.7). The idea is that these are tangible memories that visitors take home, keeping their experience and interest in the Flebocollect programme alive, in addition to stimulating a sense of belonging to the project. Further, by wearing the T-shirts and tags, visitors will contribute to the initiative's visibility and reach.

The field of citizen science has been growing, especially in recent decades (Vohland et al., 2021). However, this way of obtaining data has not always been well accepted by the scientific community, likely because of today's more executive and results-oriented culture (Haklay, 2013; Vohland et al., 2021; Yanai & Lercher, 2023). Because of this, the publication of scientific articles and book chapters along with

Fig. 10.6 Personalized T-shirts for project participants

Fig. 10.7 Junior Entomologist name tags for children

scientific events also play a key role in disseminating and understanding the project as scientific research. The use of formal recognized means of sharing research results and discoveries with the scientific community guarantees the quality and validity of the information presented. Above all, academic events bring together researchers and enthusiasts from different areas, providing an environment conducive to the exchange of knowledge and the formation of new partnerships and, consequently, expanding opportunities for research and impact. Thus, the dissemination stage of our project has proven essential to opening minds to new concepts, introducing young scientists to more exploratory and creative ways of doing science and non-specialized individuals to interesting new challenges.

Our project has also had an international impact. The Flebocollect initiative has attracted the collaboration interests of researchers and educators from countries such as Brazil, Colombia, Sri Lanka, Cape Verde (Caballero Méndez et al., 2023) and Uganda, making it clear that the demand for innovative solutions to reduce the impact of infectious diseases through the sustainable control of vectors goes beyond national borders. Collaboration between countries is one of the objectives of the project and could be key to dealing with neglected diseases in terms of sharing knowledge, resources, and best practices among nations (Gresham et al., 2013; Gebreyes et al., 2014).

In creating the road map to fight neglected tropical diseases, the World Health Organization consider the alignment of operating models and culture a fundamental pillar that supports global efforts to control, eliminate and, in due time, eradicate neglected tropical diseases such as leishmaniosis. The ultimate aim is to understand how a global strategy model can be adapted to each region, also reaching local levels, and facilitating its assimilation. However, implementing global strategies in regions with great contrasts in relation to their sociopolitical and economic realities represents a great challenge, especially when considering that these diseases mostly affect the most vulnerable populations, are often neglected due to lack of resources and inadequate infrastructure, and are most of the times limited health priorities (WHO, 2021).

In so-called developed countries, citizen science initiatives are facilitated by their social structure (Vohland et al., 2021). In regions with favourable social and economic contexts, there is much more investment in research and education as well as access to this type of projects. In countries like Spain, for instance, when considering the natural history of leishmaniosis, poverty is not associated with its epidemiology (de Freitas Milagres et al., 2023). In contrast, in other parts of the world, poor living conditions have been strongly linked to cases of leishmaniosis (Okwor & Uzonna, 2016).

Luckily, citizen science has shown the potential to discover new solutions even in the face of complex scenarios such as poverty, health inequalities and climate change (Ashepet et al., 2021; Hodgkinson et al., 2022). This is because this field of investigation appears as an alternative to traditional decision-making, involving integrative thinking between organizations, domains and sectors and, in some cases, between countries (Hodgkinson et al., 2022). As a participatory and inclusive practice, citizen science can be seen as an opportunity to engage diverse social groups, including marginalized ones, promoting sustainable development (Varga et al., 2023).

Currently, the implementation of the Flebocollect project in other countries is still in its initial stages, always adopting a critical and inclusive stance. By creating connections and disseminating its philosophy in other areas, the Flebocollect initiative reaffirms its responsibility to involve populations threatened by leishmaniosis and other infectious diseases, in favour of the development and autonomy, using this knowledge to mitigate the production of inequality and dependence (Varga et al., 2023). Hence, for the ethical and successful implementation of initiatives such as Flebocollect in different social, cultural, and economic contexts, it is essential to always add a critical dimension to the work, viewing its health consequences within the broader political scenario.

Thus, the Flebocollect project's impact extends beyond scientific realms, bearing significant social and political consequences. By embracing citizen science principles, it challenges conventional scientific practices and fosters greater public engagement with research initiatives. Through its national dissemination efforts, including workshops, social media presence, and participation in science fairs, Flebocollect has democratized scientific knowledge, making it accessible and inclusive. This outreach not only stimulates curiosity and interest in science among the general populace but also cultivates a sense of ownership and belonging within the community. Furthermore, by attracting international collaboration, Flebocollect transcends geographical boundaries, emphasizing the global relevance of addressing neglected tropical diseases. However, it's crucial to acknowledge the sociopolitical complexities inherent in implementing such initiatives, particularly in regions marked by economic disparities and healthcare challenges.

10.5 Results of the Flebocollect Citizen Science Project

The results of the project are still being analyzed, but preliminary results derived from the 841 satisfaction surveys received can be presented. As it was exposed before, the aim of the project was to explore the following dimensions: (1) Participants perception about citizen science, (2) The project's impact in their science perception, (3) Participants future action in citizen science projects and (4) Personal assessment of the project.

Concerning the first dimension *"Participants perception about citizen science"*, after participating in the whole projects 82% of the citizens were aware of the "citizen science" concept. In addition, when participants were asked in which topics of citizen science projects they would like to participate in the future, the answers were very varied. The top 5 most mentioned topics were lab experiments, health and sickness, sports and healthy habits, biodiversity, and new technologies. Other stated topics were renewable energy, nutrition, prototype construction and space and astronomy (Fig. 10.8).

In relation to the second dimension, *"The project' impact in their science perception"*, participants were asked about their auto perception once they had experienced the whole project. A Likert scale was chosen in order to measure the agreement degree to several statements: (1) "My participation in this project has made me feel valuable and with something to contribute to the scientific world" (ST- 1), (2) "My participation in this project has made me feel more motivated in science class o towards science events" (ST-2), (3) "I have learned things that I will not forget thanks to my participation in the project" (ST-3) and (4) "My participation in this

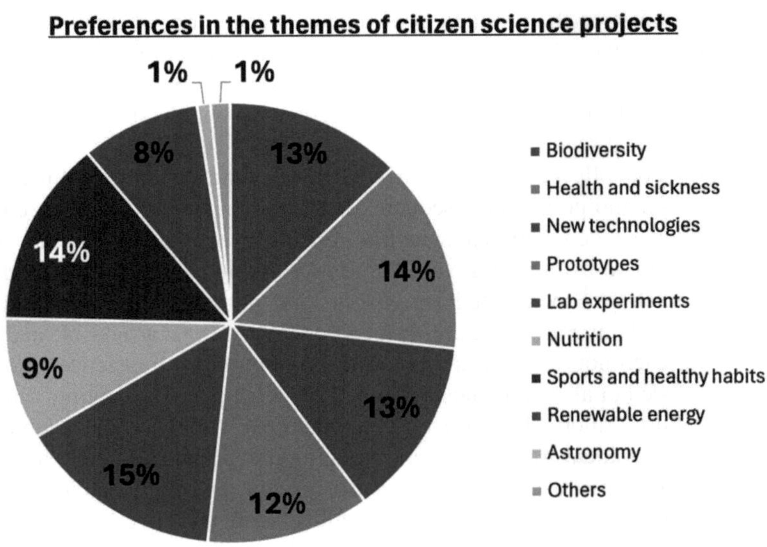

Fig. 10.8 Participants' preference for citizen science projects' themes

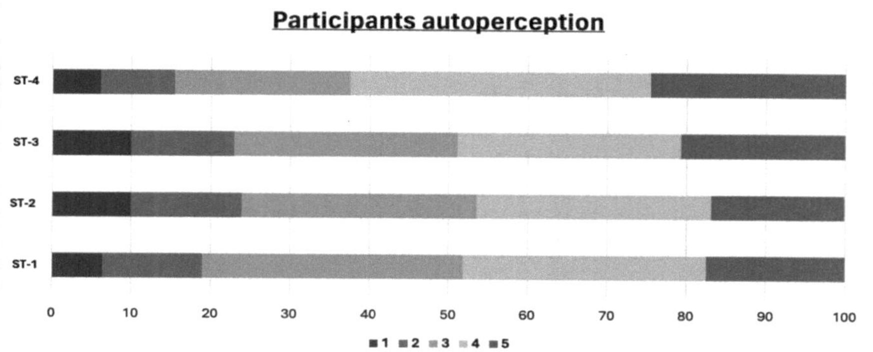

Fig. 10.9 Participants results in the Likert scale test that measured their agreement level towards 4 statements related to the projects' impact in their science perception

project has made me feel that not only scientists can do science" (ST-4). Results are exposed in Fig. 10.9.

Referring to *"Participants future action in citizen science projects"*, when participants were asked for future participation more than 78% stated that they would like to take part of more citizen science projects. Between the reasons exposed, two options stand out above the rest: their participation in this project has significantly increased their interest in science (79,6%) and they have learned numerous things without studying, just by developing the dynamic activities proposed (77,9%).

Lastly, in the *"Personal assessment of the project"* dimension, participants had to evaluate the two main activities that formed the project: the board game and the traps construction and installation. Both obtained a similar punctuation (1 to 10 scale): the board game was given an 8,66 and the traps DIY workshop an 8,78. Participants also proposed some improvements like using more times for the traps building activity (repeat the project the following years), the design of more board games about other disease's outbreaks and an additional activity for insect classification rather than the given guide.

10.6 Threats and Future Perspectives of a Citizen Science Project

As previously seen, citizen science projects have the intrinsic characteristic of always evolving, with new ideas and questions being constantly formulated by teachers, citizen participants and scientists alike. Nevertheless, any such project has potential risks and threats that could jeopardize the integrity of any initiative. In consequence, it is always necessary to focus on the main objectives of the project, and evaluate what are its potential capacities, human and material, at each specific stage of the life cycle of the research group.

Several risks always need to be taken into consideration before recruiting any citizen science collective. Based on our experience during the development of the Flebocollect initiative, three main risk categories have been defined: economic risks, technical risks, and human risks.

The main initial problem in any citizen science project is most likely related to economic limitations. A financial source, such as a grant or funding, is always recommended to establish any citizen science team. For our project, funds from the Bringing Research Into ThE Classroom initiative (BRITEC), financed by the Erasmus+ KA2 Programme, were used to organize several meetings and purchase materials to construct the first trap prototypes. Currently, due to a following funding source, the Spanish Foundation for Science and Technology (reference number: FCT-21-16782) whose goal was to implement educational activities in high schools for the promotion of health education, new educational projects were being conducted and/or assessed. This grant has allowed us to finance the activities previously described in schools of the Community of Madrid and other regions. However, it is important to note that most citizen science projects have limited financing sources, both in terms of budget and time, and this may determine an early end of any citizen science group, once this funding has been withdrawn. Hence, secondary funding sources, whenever possible, are always recommended for the follow up of any such project.

Secondly, technical risks need to be taken into consideration. One of the limitations of any citizen science project is the lack of technical knowledge in a specific field. In the case of the Flebocollect project, the original main objective was the construction and testing of a new DIY light trap designed to capture phlebotomine sand flies. Knowledge on both the biology of sand flies and circuit design were necessary for the conceptualization and development of these traps. The technical risks of any project can be overcome by setting up a multidisciplinary work team.

Thirdly, and most importantly, the main risks of any citizen science project are human risks. Among those, we defined two subcategories: human risks associated with occasional participants of the project and those related to the project itself. First, if the project cannot reach its main audience and no interest is raised among the selected communities, no citizen participation will be achieved. As a solution, social media and, if possible, raising awareness of the activities conducted through other media (e.g., fliers, television, radio) is a great way of sparking interest in any citizen science project (Alarcón-Elbal et al., 2024). It is also important to target projects aimed at solving problems that are close to the reality of citizens. Timing is another factor to consider. Flebocollect emerged at a time coinciding with the recent largest outbreak of human leishmaniosis recorded to date in Madrid. Since the COVID-19 pandemic, people are much more concerned about the emergence of new diseases, and this topic is currently trending. Also, one of the key threats for any citizen science project is the loss of interest among the main staff in charge of conducting the project. It is important to note that most participants of these projects are unpaid volunteers, with no incentive other than knowledge or experience acquired during the course of the project. This determines a risk that members of the research team could drop out of the project at any time. While newcomers may be especially interested in a citizen science project, with time, factors such as fatigue and tasks becoming tedious may come into play and be a cause of drop outs. So, it is important to somehow carefully select candidates by finding out their interests and why they decided to participate and also to maintain their interest once selected by giving them fulfilling realistic tasks, and by encouraging and acknowledging their work in every possible way (Asingizwe et al., 2020).

In conclusion, it is always necessary to understand the strengths and weaknesses of working with a citizen science team at any given time. A solid funding source, the construction of a multidisciplinary work team, the identification of the main target audience and maintaining the project's staff interested in the project is necessary for the long-term establishment of any citizen science team.

Currently, the objective of the Flebocollect team is to improve the effectiveness of the DIY traps by assessing new strategies to capture phlebotomine sand flies, and to further investigate their effectiveness for capturing other insects of public health interest (e.g., Culicidae mosquitoes, biting midges). We are also currently considering the implementation of these tools as mass trapping devices nation-wide in schools across Spain with the dual objective of building a citizen science-powered phlebotomine sand fly distribution map and improving children's perceptions of science and its role in society.

The findings of this research emphasize the promising advantages of citizen science education initiatives in schools, addressing knowledge gaps, and fostering progress in both research and public awareness. Despite the obstacles, school citizen science holds significant potential to transform research, providing deeper insights into this from both scientific and societal viewpoints.

Acknowledgments This study was funded in 2022–2023 academic course by the Spanish Foundation for Science and Technology—Ministry of Science and Innovation, to carry out activities in the field for the promotion of scientific, technological and innovation culture (Id. BDNS n° 592032) through the project "Flebocollect: Didactic strategy for the development of scientific competence through the study of emerging diseases" reference FCT-21-16782. We also wish to thank the teachers and students of the 20 participating high schools for their interest and enthusiasm (**Madrid**: IES Sapere Aude, IES Infanta Elena, CEIPSO Padre Garralda, Escuela Ideo, IES Camilo Jose Cela, IES Joaquín Rodrigo, IES San Agustín de Guadalix, IES Jorge Guillen, IES Carpe Diem, CEIPSO El Encinar, CEIP Pablo Picasso, IES Alameda de Osuna, Colegio Santo Domingo, IES Juan de Mairena; **Guadalajara**: IES Liceo Caracense; **Aragón**: Centro de interpretación de la Naturaleza de la Reserva Natural de los Sotos y Galachos del Ebro, IES San Alberto Magno; **Granada**: IES Alhendín; **Alicante**: IES Nou Derramador de Ibi; **Málaga**: IES Mare Nostrum).

References

Alarcón-Elbal, P. M., López-de-Felipe, M., Gil-Torró, I., García, I., Mateo, P., & Bueno, R. (2024). Knowledge, attitude, and practices of adults and children towards the Asian tiger mosquito, *Aedes albopictus* (Diptera: Culicidae), in a recently invaded municipality of Valencia, Spain. *International Journal of Tropical Insect Science, 44*(4), 1633–1648.

Alexander, B. (2000). Sampling methods for phlebotomine sandflies. *Medical and Veterinary Entomology, 14*(2), 109–122.

Alidosti, M., Shahnazi, H., Heidari, Z., & Zamani-Alavijeh, F. (2022). Design and evaluation of two educational media in the form of animation and games to promote the cutaneous leishmaniasis prevention behaviors in adolescent female. *BMC Public Health, 22*(1), 1–10.

Alvar, J., Yactayo, S., & Bern, C. (2006). Leishmaniasis and poverty. *Trends in Parasitology, 22*, 12.

Amelia, V. L., & Setiawan, A. (2019). El juego de mesa como medio educativo para el conocimiento sobre la prevención del dengue en niños en edad escolar. *Enfermería Global, 18*(56), 253–272.

Aranda, C., Martínez, M. J., Montalvo, T., Eritja, R., Navero-Castillejos, J., Herreros, E., Marqués, E., Escosa, R., Corbella, I., Bigas, E., Picart, L., Jane, M., Barrabeig, I., Torner, N., Talavera, S., Vázquez, A., Sánchez-Seco, M. P., & Busquets, N. (2018). Arbovirus surveillance: First dengue virus detection in local Aedes albopictus mosquitoes in Europe, Catalonia, Spain, 2015. *Euro Surveillance, 23*(47), 1700837.

Arce, A., Estirado, A., Ordobás, M., Sevilla, S., García, N., Moratilla, L., de la Fuente, S., Martínez, A. M., Pérez, A. M., Aránguez, E., Iriso, A., Sevillano, O., Bernal, J., & Vilas, F. (2013). Re-emergence of leishmaniasis in Spain: Community outbreak in Madrid, Spain, 2009 to 2012. *Euro Surveillance, 18*(30), 20546.

Ashepet, M. G., Jacobs, L., Van Oudheusden, M., & Huyse, T. (2021). Wicked solution for wicked problems: Citizen science for vector-borne disease control in Africa. *Trends in Parasitology, 37*(2), 93–96.

Asingizwe, D., Poortvliet, P. M., Koenraadt, C. J. M., van Vliet, A. J. H., Ingabire, C. M., Mutesa, L., & Leeuwis, C. (2020). Why (not) participate in citizen science? Motivational factors and barriers to participate in a citizen science program for malaria control in Rwanda. *PLoS One, 15*(8), e0237396.

Bermejo, A., & Gálvez, R. (2023). OUTBREAK: El misterioso caso del brote de leishmaniosis de la Comunidad de Madrid. In M. G. M. de Espinosa & A. H. Sánchez (Eds.), *Experiencias y estrategias de innovación educativa en ciencia, tecnología, ingeniería y matemáticas* (Vol. III, pp. 213–220). Editorial SM.

Bonney, R., Phillips, T. B., Ballard, H. L., & Enck, J. W. (2016). Can citizen science enhance public understanding of science? *Public Understanding of Science, 25*(1), 2–16.

Butala, C., Fyfe, J., & Welburn, S. C. (2021). The contribution of community health education to sustainable control of the neglected zoonotic diseases. *Frontiers in Public Health, 9*, 729973.

Caballero Méndez, E., Ferrero Gómez, L., Fernández Serafín, C., López de Felipe, M., Bermejo, A., Ribeiro Rocha, H. D., & Gálvez, R. (2023, May 26). Adaptação da armadilha de luz DIY na macaronésia como estratégia inovadora e sustentável para o combate de vetores causadores de doenças [Conference presentation]. In *2ª edição Simpósio de investigação científica da Universidade Jean Piaget de Cabo Verde*, Praia, Cabo Verde.

Cilleruelo, L., & Zubiaga, A. (2014). Una aproximación a la educación STEAM. Prácticas educativas en la encrucijada arte, ciencia y tecnología. In J. Cruz & M. Díaz (Eds.), *Investigar en psicopedagogía, una realidad en auge* (pp. 22–38). Servicio Editorial de la Universidad del País Vasco.

Clegg, A. (1979). Craftsmen and the origin of science. *Science & Society, 43*(2), 186–201.

Collantes, F., Delacour, S., Delgado, J. A., Bengoa, M., Torrell-Sorio, A., Guinea, H., Ruiz, S., Lucientes, J., & Alert, M. (2016). Updating the known distribution of *Aedes albopictus* (Skuse, 1894) in Spain 2015. *Acta Tropica, 164*, 64–68.

Cosme, L., Jr., Turchen, L. M., & Guedes, R. N. C. (2020). Insect world: Game-based learning as a strategy for teaching entomology. *The American Biology Teacher, 82*(4), 210–215.

Couso, D. (2017). Per a què estem aSTEM?Un intent de definir l'alfabetització STEM per a tothom i amb valors. *Ciències, 34*, 22–30.

Da Silva Santos, L. R., Souza-Pinho, M. J., de Jesus, M. S., & Kalil, A. (2022). Educational games in the high school: Implicate future teachers in the pursuit for new teaching strategies. *IJERI: International Journal of Educational Research and Innovation, 17*, 27–44.

Danilina, Y. V. (2022). Science popularization as an element of innovative communications. *Scientific and Technical Information Processing, 49*, 21–29.

de Freitas Milagres, T., López-de-Felipe, M., da Silva, W. J., Martín-Martín, I., Gálvez, R., & da Silva, O. S. (2023). Same parasite, different outcomes: Unraveling the epidemiology of *Leishmania infantum* infection in Brazil and Spain. *Trends in Parasitology, 39*(9), 774–785.

de Freitas Milagres, T., Bermejo, A., López, de Felipe, M., Santos da Silva, O., Mora-Urda, A.I., & Gálvez, R. (2025). Thinking outside the box: Citizen science to monitor phlebotomine sand flies in Spain. *Acta Tropica, 265*, 107619.

De la Vega, G. J., Falconaro, A. C., Soria, L., & Corley, J. C. (2022). Integrated pest management education: A video-game to improve management of *Drosophila suzukii*, soft-skin fruit pest. *Neotropical Entomology, 51*(5), 801–807.

Eritja, R., Ruiz-Arrondo, I., Delacour-Estrella, S., Schaffner, F., Alvarez-Chachero, J., Bengoa, M., Puig, M. A., Melero-Alcibar, R., Oltra, A., & Bartumeus, F. (2019). First detection of *Aedes japonicus* in Spain: An unexpected finding triggered by citizen science. *Parasites & Vectors, 12*(1), 53.

Flebocollect App. (2023). Retrieved from https://flebocollect.com/user/

Flebocollect Web. (2023). Retrieved from https://www.flebocollect.com

FlebocollectIG [@flebocollect]. (n.d.). Instagram. Retrieved July 7, 2023, from https://www.instagram.com/flebocollect/

FlebocollectX [@flebocollect]. (n.d.). X. Retrieved July 7, 2023, from https://x.com/flebocollect

Gálvez, R., Gollerizo, A., & Clemente, M. (2019, July 15–19). Harnessing student citizen science in and around the classroom to study sand fly populations in Madrid (Spain) [Conference presentation]. In *10th International Symposium On Phlebotomine Sandflies*, Isla De San Cristobal, Ecuador.

Gálvez, R., Montoya, A., Cruz, I., Fernández, C., Martín, O., Checa, R., Chicharro, C., Miguelanez, S., Marino, V., & Miró, G. (2020). Latest trends in *Leishmania infantum* infection in dogs in Spain, part I: Mapped seroprevalence and sand fly distributions. *Parasites & Vectors, 13*(1), 204.

Gálvez, R., Bermejo, A., & López de Felipe, M. (2022a, September 19–23). Flebocollect DIY light traps: Making of a citizen science project for schools and researchers to study phlebotomine sand flies. In *Conference presentation at the symposium "Engaging the public in vector biology research and vector-borne disease control". SOVE 8th International Congress*, Honolulu, Hawaii.

Gálvez, R., López de Felipe, M., & Yebes, F. (2022b). Citizen science set in motion: DIY light traps for phlebotomine sand flies. *Preventive Veterinary Medicine, 200*, 105589.

Gebreyes, W. A., Dupouy-Camet, J., Newport, M. J., Oliveira, C. J., Schlesinger, L. S., Saif, Y. M., Kariuki, S., Saif, L. J., Saville, W., Wittum, T., Hoet, A., Quessy, S., Kazwala, R., Tekola, B., Shryock, T., Bisesi, M., Patchanee, P., Boonmar, S., & King, L. J. (2014). The global one health paradigm: Challenges and opportunities for tackling infectious diseases at the human, animal, and environment Interface in low-resource settings. *PLoS Neglected Tropical Diseases, 8*(11), e3257.

Gil, I., Flor, A., Alarcón-Elbal, P. M., López de Felipe, M., García, I., Mateo, P., & Bueno, R. (2022, November 28–29). Quality assessment of mosquito mass trapping implementation and its intrinsic efficacy in a field study for *Aedes albopictus* control. In *Conference presentation. Xth EMCA workshop*, Mendrisio, Switzerland.

Gresham, L., Smolinski, M., Suphanchaimat, R., Kimball, A. M., & Wibulpolprasert, S. (2013). Creating a global dialogue on infectious disease surveillance: Connecting organizations for regional disease surveillance (CORDS). *Emerging Health Threats Journal, 6*, 19912.

Hadjichambis, A., Paraskeva-Hadjichambi, D., Georgiou, Y., & Adamou, A. (2023). How can we transform citizens into 'environmental agents of change'? Towards the citizen science for environmental citizenship (CS4EC) theoretical framework based on a meta-synthesis approach. *International Journal of Science Education, Part B*, 1–21.

Haklay, M. (2013). Citizen science and volunteered geographic information: Overview and typology of participation. In D. Sui, S. Elwood, & M. GoodchildIn (Eds.), *Crowdsourcing geographic knowledge* (pp. 105–122). Springer.

Hodgkinson, I. R., Mousavi, S., & Hughes, P. (2022). New development: Citizen science – Discovering (new) solutions to wicked problems. *Public Money & Management, 42*(2), 133–136.
Huber, B., Barnidge, M., Gil De Zúñiga, H., & Liu, J. (2019). Fostering public trust in science: The role of social media. *Public Understanding of Science, 28*(7), 759–777.
Jääskä, E., & Aaltonen, K. (2022). Teachers' experiences of using game-based learning methods in project management higher education. *Project Leadership and Society, 3*, 100041.
Kalaentzis, K., Kazilas, C., Demetriou, J., Koutsoukos, E., Avtzis, D. N., & Georgiadis, C. (2021). Alientoma, a dynamic database for alien insects in Greece and its use by citizen scientists in mapping alien species. *Insects, 12*(12), 1101.
Khelifa, R., & Mahdjoub, H. (2021). EcoDragons: A game for environmental education and public outreach. *Insects, 12*(9), 776.
Kolb, A. Y., & Kolb, D. A. (2005). Learning styles and learning spaces: Enhancing experiential learning in higher education. *Academy of Management Learning & Education, 4*(2), 193–212.
Lave, E., & Wanger, J. (1991). *Situated learning: Legitimate peripheral participation*. Cambridge University Press.
Liebert, W. A. (2019). Communicative strategies of popularization of science (including science exhibitions, museums, magazines). In A. Leßmöllmann, M. Dascal, & T. Gloning (Eds.), *Science communication* (pp. 399–416). De Gruyter.
López de Felipe, M., Orán, J. P., Fernández, C., Alarcón-Elbal, P. M., Gil, I., García, I., Mateo, P., & Bueno, R. (2022, November 28–29). Conceptualization and logistics of the traps distribution phase of the citizen science project NESCOTIGER: New strategies for the control of *Aedes albopictus* in residential areas [Conference presentation]. In *Xth EMCA workshop*, Mendrisio, Switzerland.
López de Felipe, M., Pérez, E., & Gálvez, R. (2023). Color preference of *Sergentomyia minuta* (Diptera: Phlebotominae) determined using Flebocollect do it yourself light traps based on LED technology. *Parasitology Research, 122*(1), 217–226.
Lupión-Cobos, T., Lagarón, D. C., Ariza, M. R., & Domènech-Casal, J. (2023). STEM education in the Spanish context: Key features and issues. In S. M. Al-Balushi, L. Martín-Hansen, & Y. Song (Eds.), *Reforming science teacher education programs in the STEM era: International and comparative perspectives* (pp. 181–198). Springer International Publishing.
MediaLab Prado. (2021). Retrieved from https://www.medialab-matadero.es/proyectos/52051/description
Monge, S., García-Ortúzar, V., López Hernández, B., López Pérez, M. A., Delacour-Estrella, S., Sánchez-Seco, M. P., Fernández Martínez, B., García San Miguel, L., García-Fulgueiras, A., & Sierra Moros, M. J. (2020). Characterization of the first autochthonous dengue outbreak in Spain (August–September 2018). *Acta Tropica, 205*, 105402.
Okwor, I., & Uzonna, J. (2016). Social and economic burden of human Leishmaniasis. *American Journal of Tropical Medicine & Hygiene, 94*(3), 489–493.
Pérez, L. N. C. (2020). Importancia de la Educación para la Salud en currículo educativo. *Revista Electrónica de Conocimientos, Saberes y Prácticas, 3*(1), 170–118.
Phillips, D. C. (1995). The good, the bad, and the ugly: The many faces of constructivism. *Educational Research and Reviews, 24*(7), 5–12.
RD 217/2022, de 29 de marzo, por el que se establece la ordenación y las enseñanzas mínimas de la Educación Secundaria Obligatoria. Boletín Oficial del Estado, núm. 76, de 30 de marzo de 2022. Retrieved from https://www.boe.es/buscar/pdf/2022/BOE-A-2022-4975-consolidado.pdf
Savin-Baden, M., & Major, C. H. (2004). *Foundations of problem-based learning*. McGraw-Hill Education.
van Swaay, C. A., Nowicki, P., Settele, J., & Van Strien, A. J. (2008). Butterfly monitoring in Europe: Methods, applications and perspectives. *Biodiversity and Conservation, 17*, 3455–3469.
Varga, D., Doran, C., Ortega, B., & Segú Odriozola, M. (2023). How can inclusive citizen science transform the sustainable development agenda? Recommendations for a wider and more

meaningful inclusion in the Design of Citizen Science Initiatives. *Citizen Science: Theory and Practice, 8*(1), 29, 1–10.
Vilas, F., Hervás, J. C., Sevilla, S., Martínez, A., Gavín, M. O., Bernal, J., Díaz, R., Iriso, A., Sevillano, O. F., Sainz, C. E., de la Fuente, S., Arce, A., Estirado, A., Frutos, J., & y Fúster, F. (2012). Brote de Leishmaniasis en la zona suroeste de la Comunidad de Madrid: Medidas de investigación y control medioambiental. *Profesión Veterinaria, 17*(79), 6–15.
Vohland, K., Land-zandstra, A., Ceccaroni, L., Lemmens, R., Perelló, J., Ponti, M., Samson, R., & Wagenknecht, K. (2021). *The science of citizen science*. Springer.
Weingart, P., Joubert, M., & Connoway, K. (2021). Public engagement with science – Origins, motives and impact in academic literature and science policy. *PLoS One, 16*(7), e0254201.
WHO. (2021). *Ending the neglect to attain the Sustainable Development Goals: A road map for neglected tropical diseases 2021–2030*. Retrieved from https://www.who.int/publications/i/item/9789240010352
WHO & UNICEF. (2017). *Global vector control response 2017–2023*. Retrieved from https://apps.who.int/iris/bitstream/handle/10665/259205/9789241512978-eng.pdf?sequence=1
Yanai, I., & Lercher, M. J. (2023). Make science disruptive again. *Nature Biotechnology, 41*, 450–451.

Chapter 11
Community Versus Citizen Science: An Activist-Based Distinction

Yael Silver and Ayelet Shavit

11.1 Introduction

Over the past few decades, a growing movement has emerged advancing public participation in scientific research. Methodologies exhibiting different characteristics and degrees of participant involvement are now used across diverse research fields and initiatives. Public participation in scientific research is popular among citizens and scientists alike, creating new opportunities for learning and developing scientific knowledge as well as for generating social and political change. This growing field of research and practice encompasses a variety of approaches. In this chapter, we will focus on two key frameworks: citizen science and community science. We highlight that each approach is suited to different needs, purposes, and contexts, no one is inherently superior to the other.

Multiple characterizations of citizen science and community science exist, with overlapping extensions and conflated meanings. We expound here upon several existing definitions, typologies and differentiations that delineate various paradigms within the domains of citizen science and community science. The distinction we propose suggests that community science serves as a research framework that systematically focuses on engagement with issues of injustice. Our primary motivation for this article is to advance this perspective; therefore, we focus primarily on the advantages and defining characteristics of community science. While we acknowledge the significant benefits of citizen science—widely discussed in the literature—we will briefly address some of them in the next section. From these research perspectives, we distill a 'universal' versus 'particular' and 'individual' versus

Y. Silver (✉) · A. Shavit
Technion Israel Institute of Technology, Haifa, Israel

Tel-Hai College, Upper Galilee, Israel
e-mail: yael.eshed@campus.technion.ac.il; ashavit@telhai.ac.il

D. Tsybulsky, T. Tal (eds.), *Citizen Science Contributions to Biology and Environmental Education*, Contributions from Biology Education Research,
https://doi.org/10.1007/978-3-031-99298-8_11

'collective' distinction, follow through its conceptual and practical implications, and argue, therefore, that community science harbors a higher potential to mitigate and/or resist social, environmental and epistemic injustice. We substantiate our argument through an empirical examination of a case study: "Town Square Academia", which pro-actively advocates for and embodies the principles of community science in a peripheral region (Aharon-Gutman et al., 2018).[1] We will illustrate how the universal-particular perspective readily suggests conflicting action paths, which lead to different results. We end by examining some of the possible weaknesses in our position and invite our colleagues to further examine this distinction, in theory and in practice.

11.2 Review

11.2.1 What Is Citizen Science?

Science policy analyst Alan Irwin (1995) and ornithologist and participatory research organizer Richard Bonney (1996) are often credited with coining the term 'citizen science.'[2] In attempting to describe projects in which nonscientists contributed scientific data, Bonney suggested the term citizen science, broadly defined as initiatives promoting public participation in scientific research (Bonney, 1996; Shirk et al., 2012; Strasser et al., 2019). Tsybulsky and Tal's entire edited book is focused on 'citizen science', hence this chapter will not delve into its various definitions.

The concept of citizen science is integral to movements towards a more open or democratized science (Dickel & Franzen, 2016; Kitcher, 2001; Peters & Besley, 2019; Smallman, 2018; Stilgoe et al., 2014; Strasser et al., 2019). These movements call for increased accessibility to and inclusion of nonprofessional scientists in the research process, building upon the premise that non-experts can make tangible contributions to scientific research. This approach empowers citizens to act in scientific arenas relevant to their lives and aligns scientific research with their needs and values (e.g., Butkevičienė et al., 2021; Krings et al., 2019).

[1] 'Periphery' and 'Center' are not only a matter of geography—though our particular case study is also situated at the difficult geographical border of Israel, Lebanon and Syria—but, perhaps more so, a matter of socio-economy. Aharon-Gutman, Schaap and Lederman (2018) explain: "These concepts have established a foothold in Israeli academia and public discourse as an effective means of characterizing inequality in the country. Indeed, the concepts of 'center' and 'periphery' have constituted a basis for movements of social change, which have placed questions regarding the division of resources on the public agenda" (p. 43).

[2] Irwin's original meaning differed in important ways from Bonney's, which became identified with the term. By the middle of the second decade of the twenty-first century, the term citizen science had become popular, with little recognition that the phrase unintentionally co-opted Irwin's original intent (for details see Cooper et al., 2016).

The most democratic vision of citizen science is associated with co-created science (Shirk et al., 2012), in which citizens take part in most or all of the research stages, such as articulating the research question, study design, data analysis, dissemination of findings, and implementation of conclusions. In such projects, citizens not only co-work with professional scientists during all research stages, but may also take full responsibility over the research in order to better address their community's interests.

Despite the multitude of projects that involve the public in the scientific process, it appears that many researchers and environmental education professionals are unaware of the various possibilities of joint research with the public and the possible contribution of this type of research to science, society and the environment. In fact, over 90% of citizen science research belongs to one action model, named 'contributory research' by Bonney et al. (2009), 'assisted research' by Roy et al. (2012) or 'crowdsourcing' according to Haklay (2012). It involves citizens only in data collection, whereas a variety of practices involve the public in additional research phases. Yet, the same terms—'democracy,' 'accessibility,' 'inclusion,' and 'diversity'—along with metaphors such as—'openness,' 'building bridges,' or 'empowerment'—are employed both in cases of extensive co-creation and where citizens have very little control over the research (Cooper et al., 2016; Strasser et al., 2019), effectively social-washing[3] the majority of science-society interactions. For example, the above terms are often used in those large projects of crowdsourcing, whereby citizens function as not much more than self-operating data collection instruments (Shirk et al., 2012).

Dillon et al. (2016) suggest a different framework, envisioning citizen science "as a mechanism for enabling the convergence of science and society" (p. 450). Their schema is, indeed, helpful in tracking theory and practice hand in hand. The authors mark three different kinds of research problems—simple, complex, and wicked—attached to different research motivations—science-driven, policy-driven, and transition-driven—practically focused on different research domains—data and information, understanding and engagement, and meaning and transformation—respectively. They place this three-option configuration on two orthogonal axes: from open, emergent to closed, prescribed research goals; and from authoritative, predetermined to participatory, co-created research interactions. Using this schema, Dillon et al. distinguish 'citizen science' from 'civic science' ('community science' in our terms) and "argue for a new model of collaborative research among scientists, educators, and the public to strengthen links between science and society with a focus on place and identity" (2016, p. 450) for community science. We could not

[3] 'Social-washing' denotes an incident where one (individual or organization) explicitly claims or implicitly builds a public image of acting for social or environmental justice while not actually doing so.

We are not making a general claim that anyone who does not distinguish between community science and citizen science is committing 'social-washing.' Rather, our claim is highly specific: we contend that the lack of distinction between these two frameworks enables and reinforces such practices by those who do commit 'social-washing.'

agree more with such a transition-driven model for science-society interaction. Yet, we wish to extend this insight to additional cases and more transformative examples. For that purpose, we suggest a community science model.

While Dillon et al. (2016) offer a single example of civic science, Jordan et al. (2016) offer an innovative, useful, and potentially online tool for a collaborative decision-making process that may produce local change. The system is built upon iterative sets of remote learning meetings within and between heterogenic groups, embedded in feedback loops to add new data, new perspectives, and corrected goals and standards. Indeed, an important and potentially transformative step forward. However, the system itself was built by experts rather than co-created, does not focus on the general public but on stakeholders, does not recognize local expertise but teaches locals how to operate the system, does not include a live—and in that sense open and emergent—dialogue among participants, and did not appear to sustain public participation across time (Jordan et al., 2022). To reach these objectives, we present below a transformative project named "Town Square Academia."

Finally, a few conceptual ambiguities are noteworthy. For example, Dillon et al. (2016) assume that science-driven citizen science tackles simple research problems focused on data and information. However, nearly anyone who follows biologists executing long-term surveys, as one of us did (Shavit & Griesemer, 2009, 2010), knows that the problems scientists tackle are far from simple (at least according to the characterization of 'simple' found in Dillon et al. (2016)). In particular, long-term survey data and metadata collection, organization, use and reuse is anything *but* simple. Bottom line, the inner connection between kinds of problems and kinds of research is difficult to justify.

Citizen science offers advantages across several dimensions, making it a powerful approach for large-scale scientific inquiry. Citizen science enables the collection of vast amounts of data, particularly in fields that require extensive sampling beyond the reach of traditional research methods (Kelling et al., 2015). For example, the eBird project has utilized citizen science data to estimate species population trends, addressing information gaps in regions lacking rigorous monitoring data (Fink et al., 2023). Similarly, the Floating Forests project demonstrated that data collected by citizen scientists can achieve accuracy comparable to expert classifications, thereby enhancing large-scale ecological research (Rosenthal et al., 2018). Its ability to engage a wide range of participants fosters broad public involvement (Silvertown, 2009), integrating diverse perspectives into the research process. Additionally, citizen science upholds scientific rigor and standardization, ensuring the reliability of findings in disciplines that demand universal sampling and stringent quality control. Rosenthal et al. (2018) demonstrated that citizen scientists can generate data comparable in quality to expert classifications, facilitating large-scale ecological research. Furthermore, by addressing globally relevant research questions, citizen science produces knowledge that transcends local contexts, making its insights applicable to multiple communities and scientific domains (Dillon et al., 2016; Shirk et al., 2012).

11.2.2 What Is Community Science?

There are many definitions of community science (Berkes & Folke, 1998; Berkes & Ross, 2013; Charles et al., 2020; Wandersman, 2003; Wilderman et al., 2003). Charles et al. (2020) defined it as "scientific research and monitoring, based on scientific modes of inquiry, which are (i) community-driven and community controlled, (ii) characterized by place-based knowledge and social learning, collective action and empowerment, and (iii) with the normative aim to negotiate, improve and/or transform governance for stewardship and social-ecological sustainability" (p. 77). According to Weisberg et al. (2023, p. 1), "Community science involves scientists and community members co-creating and co-executing scientific research" and notably addresses "historical and ongoing inequities between scientific professionals and the communities where scientific investigation takes place" (Weisberg, 2021, p. 7). Community science endeavors to improve community settings, fostering the capacity, resilience, and overall well-being of individuals within these communities, while also striving to deter detrimental behaviors at both the individual and community levels. A unifying feature across community science definitions is the notable emphasis on the community's pivotal role and function within scientific research, underscored by the collaborative relationships forged with the community as an integral component of the research process.

We want it to be clear that the terms science and scientific research are used here according to "actor speech." The actors, in our case, are active practitioners of any kind of systematic collection and analysis that produce empirically testable results (broadly construed). We emphasize that this is not a general definition of 'science' nor a concept confined to the 'natural sciences' alone. It can include types of information and modes of professionalized inquiry that may be classified within the social sciences and humanities. Indeed, the attempt to characterize a logical (or otherwise rational) "principle of demarcation" between science and pseudo-science has historically failed (Kuhn, 1962). In fact, we use the word science here somewhat reluctantly, as all too often when referring to 'science' in every-day speech the human dimension is left out and the humanities and social sciences disciplines become marginalized relative to natural sciences. This bias can occur even if those using the term are themselves educators, historians or psychologists and even if they explicitly indicate that *all* forms of knowledge are included in their meaning of 'science' (Charles et al., 2020).

11.3 Typology of Public Participation in Science

Projects within the domains of citizen science and community science are classified in the literature based on multiple different criteria, including, the extent of public participation, the degree of public influence, the nature of the activities involved, and the activity's geographical scope (Roy et al., 2012). Level of public engagement (also called participation or involvement) is a common method of categorization (Boersch-Supan et al., 2019; Bonney et al., 2009; Shirk et al., 2012). For both citizen science and community science, scholars have proposed a spectrum of increasing levels of public involvement. At the one end of this spectrum, scientists conceive the research, perform data analysis and formulate conclusions, whereas citizens primarily collect data. At the other end. Scientists and citizens co-create comprehensive collaboration.

Another sorting system deals with the method of citizen engagement, including its degree of physical activity, from online to physical (Haklay, 2015). Haklay (2012) describes a four-level hierarchy. In the first level, "crowdsourcing," people participate as sensors—not knowing how the data contributes to addressing scientific questions. In the second level, "distributed intelligence," participants carry out cognitive tasks for data interpretation. In the third level, "participatory science," participants contribute to the problem definition and carry out data collection tasks.

Finally, in the fourth level, "extreme citizen science," participants are deeply involved in the research process through co-creation, which supports collaboratively identifying the problem, forming the research questions, designing the tools and methods to support data collection, and collecting and analyzing the data.

A third sorting approach to participation (Haklay et al., 2021) classifies community and citizen science projects based on their geographic and policy influence. The geographic dimension marks research on local, urban, regional, state, continental, or global scales. These different scales have notable political implications for policy formulation, research oversight, patent rights, partner involvement and communication methods. Both citizen and community science projects may span any of these scales, yet citizen science typically aims for global influence and community science for local or urban influence (yet mostly depends upon the specific geographical delineation of the community). The dimension of policy pertains to the sphere of influence of both citizen science and community science. It encompasses diverse domains such as environmental decision-making, urban planning, agricultural and food systems, humanitarian aid and development, scientific awareness, and support for scientific endeavors.

Finally, Charles et al. (2020) have classified projects based on the question of "Who owns the knowledge" for all phases, from research planning to the result or patent registration phase in some cases. At one pole, universities, colleges, or private or government research institutes own the knowledge produced as a result of their interaction with the community. This end remains academia owned, regardless of who initiated the participatory interaction: whether it was scientists with a question in need of public help or it was a "science shop" project where nonscientists pose a problem in need of an answer (Urias et al., 2020). At the other pole of knowledge ownership stand local communities that rely on expertise within their own ranks, and if engaging with academia it is to make that knowledge publicly visible and/or protect it from theft or misuse (Mgbeoji, 2007). A third option, is dual ownership. It requires a dialogic process of collaboration in which communities and scientists sign a contract—officially or unofficially –to conduct research or supply necessary data, explicitly discussing and reaching an agreement on the scope and level of knowledge ownership, somewhat similar to other contracted projects or community consultation.

Citizen science encompasses a wide range of research projects, including projects particular or localized to certain areas or communities. That said, this paper offers a categorical distinction—not a gradual continuum[4]—between community and citizen science. For a project to be characterized as 'community science', the following practical commitments are all simultaneously necessary: a pre-disposition to notice *locality*, take part in a *particular* entity, namely a *collective* (a community or community of communities) and to *pro-actively* labor for that entity's advancement. For citizen science projects, activism, although plausible, is not a necessary disposition. Other practical commitments are necessary. For example, an inclination to seek and provide *universality*—of the research data, explanation, implication etc., and to treat whichever entity is involved as a collection of distinct individuals. Therefore, we argue that even if a project is more physically or culturally localized than average, if it abides by our categorical distinctions, i.e., aims for universality rather than particularity and to attend individuals rather than collectives, it is better described as citizen science.

11.4 The Relation Between 'Citizen Science' and 'Community Science'

There are many different ways to describe the relations between the concepts of citizen and community science. One explanation for this literary phenomenon is that many definitions of 'citizen science' are extremely broad. There are notable researchers in the field who understand 'community science' as a special kind of 'citizen science', involving the community more integrally and with a higher impact on the research (Bonney et al., 2016; Dillon et al., 2016; Haklay, 2015; NAS-EM, 2018, Strasser et al., 2019). Therefore, these researchers and their teams suggest interesting and very useful mapping of citizen science but invest relatively little in distinguishing it from community science.

A clear and helpful start for such a distinction is Strasser et al. (2019). They carefully delineate—currently and historically—the main epistemic practices used by citizen science, noting that such a typology "...draws attention to practices not carried out under the banner of "citizen science," such as "participatory action research" and "community-based research," but that might nevertheless be essential to understanding public participation in the production of scientific knowledge" (p. 58). We agree that these central practices are better described under a different banner, that of community science, and attempt to conceptually clarify and empirically illustrate *why* this is the case.

---

[4] Please note that the continuous axis in Table 11.1 refers to the democratization of science, not to the community-citizen distinction.

Other researchers have also referred to community science as distinct from citizen science (Wandersman, 2003; Wilderman et al., 2003), and this view has been gaining momentum recently (Charles et al., 2020; Weisberg et al., 2023). The distinction in the literature today is mainly based upon different goals for these endeavors (expanding scientific knowledge versus promoting a community; Charles et al., 2020) and the identity of the research leaders (professional versus non-professional heads of study; Haklay, 2015; Shirk et al., 2012).

A pivotal practical distinction between citizen science and community science revolves around the role of the primary planner and leader of research activities. Conceptually, there need not be a *prima facia* distinction, since in both types of interaction, citizens and scientists cooperate. However, in practice, in most citizen science projects professional academic scientists take on the responsibility of planning and leading the research. Conversely, in community science, a much larger portion of projects are conceived and led by non-professional scientists, often by the community itself (Schrögel & Kolleck, 2019; Shen, 1975; Shirk et al., 2012). In the next section, we suggest a novel conceptual distinction that may begin to explain why this structural type of difference happens, and tends to repeat itself.

Wandersman (2003) highlights key attributes of community science. Community values and connections that delve into the historical, political, social and cultural facets of the community. Noticing social relationships in the community, focusing on enhancing community capacity and strengthening community infrastructure are more prevalent in community science projects. This is due to the fact that meticulous consideration and incorporation of these characteristics is imperative during collaborative research planning with community members; unlike citizen science projects wherein these characteristics may not necessarily be explicitly addressed (García et al., 2021; Shirk et al., 2012).

Why is it that working with "a citizen" on a scientific project readily suggests different practices and theoretical perspectives from working with "a community member"? In the next section, we argue for a new conceptual distinction between community and citizen science, which, we hope, may begin to answer this question.

11.5 A Novel Distinction Between Community and Citizen Science

We argue that community science and citizen science are essentially different in category. In other words, it is not that merely the degree of science-society interaction is more intense in the former, but that a different kind of interaction takes place, its description and interpretation based on different epistemic desiderata and best measured by different model variables.

Both interactions invite professional and non-professional scientists to conduct scientific research together; yet the crux of the difference is that citizen science extends a *universal*, boundless and particular-less invitation to everyone to participate as context-free, independent *individuals*; while community science focuses on a *particular community* or set of communities, to which participating individuals are linked and/or constrained. The former asks all to cast aside particular attachments and identities whereas the latter asks all to explicitly notice them, including their in-out borders. The 'universal-particular' and 'individual-collective' semantic distinctions are important, since the usage of different concepts generates different modes of perception, which readily suggest different practices of action.

'Universal' and 'particular' (as well as 'universal-particular axis') are concepts that denote various different instances (e.g., type of participants, research questions, research methodology). While it is common for a focus on a particular community to coincide with a particular research question, and vice versa, this is not always the case. We will consider the research question independently of the research participants. Nevertheless, we argue that in community science, it is essential for both the participants and the research question to be particularistic.

This distinction differs from the current literature, yet also builds upon it. For example, we also employ—as common in the literature—a continuous axis of increase/decrease of science democratization. However, unlike most, we do not perceive the community and citizen science distinction as only a matter of degree on this axis. 'Science Democratization' combines several elements: the degree of professional scientists' openness[5] to other research goals (and their derived research questions and methods), and the level of non-professional scientists' participation and co-creation in the actual research process.[6] We therefore consider 'participation', 'openness' and 'democratization' on the same single axis direction (see Table 11.1). In that sense, our table differs from Dillon et al.'s (2016) that positions 'open' and 'participatory' as orthogonal and independent axes.

Gradually moving on the democratization axis, scientific investigation increasingly opens to non-professional scientists as research partners, with two working models typically referred to 'science shops' and 'citizen science'. In the former, an individual has a particular question or problem (e.g., the taste of water at his home, his agricultural field, etc.), to be answered by an individual applicable scientist in the lab or classroom. This highly particularist type of reciprocal interaction is mostly popular for research science pedagogy and remains in current use (Urias et al., 2020). A somewhat similar reciprocal interaction takes place in citizen science projects, though more often than not the research question emerges from the scientist rather than the citizen. Either way, the default aim of citizen science is focused on universal—rather than localized questions—and its practice is intended to be—ideally—universally open for *every* individual to participate.

[5] We are helped by Dillon et al.'s (2016) contribution and vocabulary. For example, 'openness' is used synonymously with 'inclusion', 'flexibility' or 'toleration' in the literature, but we use the former.

[6] In practice, according to actor's speech, the axis of 'science democratization' is used similarly to the axes of science 'openness' and of 'participatory'. This happens because these concepts are not practically independent but rather foreground very similar goals and working practices.

Table 11.1 Universality versus particularity and individuality versus collectivity shape research interactions within and between science and society

	Collectivity	Individuality
Particularity	Community Science	Science Shop
	Axis of Science	Democratization
Universality		Citizen Science

Yet community science, alongside other activist research programs—such as Critical Theory (e.g., Freire, 1970; Giroux, 1983), feminist philosophy of science (e.g., Longino, 1990) and Standpoint Theory (e.g., Wylie, 2012)—agree that such an idealized, neutral representation of reality, with its individual-based colorblind and genderblind view of science-society interactions, is a major concern. The common ground for all these working models is the explicit argument that "science as usual" is homogenous in a way that replicates hegemonic ideology and enacts unjust stratification upon underrepresented groups. In that sense, community science is practically synonymous with Haklay's (2012) 'extreme citizen science' and Dillon et al.'s (2016) 'transformative civil science', yet we argue that only by noticing the constituting categories can one explain—at least partly—*why* and *how* the synonymity emerges.

So far, we identified different working models for science-society interaction as different positions on the axis of science democratization. We do not argue that this axis denotes an overall agreed scale of increasing democratization, since one perspectives perceived 'democracy' as opening access to a greater number of people while the other also counts these people's differences (e.g., their geographical, cultural or socio-economic heterogeneity) as equally important. Nor do we claim this democratization axis is empirically mistaken, but only that it is explanatory insufficient, glossing over causal "difference makers" in one's working model. That is, although an increasing degree of democratization fits well within our model, this axis alone neglects to notice the categories that shape the practical meaning of 'science democratization' and its resulting working practices, which we deem especially relevant. Different categories construct different modes of operation, for example direct attention to those absent from the deliberation process and the decision table, their structural equality or inequality, and how to incorporate dissent during deliberation (see details below and in Sect. 11.6).

In other words, we argue that even if the research was highly democratized, it could still be a 'citizen science' rather than a 'community science' project. Within each category one could increase or decrees its level of democratization, but one need not assume a single common democracy axis crossing them both. The CBC—Christmas Bird Count—is a useful example. It is a long-term point-count bird survey of the Audubon society, a pro-active NGO, initiated and managed to date outside academic institutions (i.e., universities and colleges). The society was established and still operates via concerned affluent individuals alongside expert naturalists, typically holding some degree of academic training yet employed, socialized and professionalized *outside* the academia: in natural history museums, governmental offices (e.g., Fish and Wildlife Service, Forest Service, Park Service), municipal or philanthropic organizations. These non-academic naturalists build the sampling protocol, which mostly takes place in near-home localities and with data-ownership and political power of usage open to all participants, thus fully *outside* the authority of professional academic science. Therefore, the CBC seems to check the list of requirements for a 'co-created citizen science' (Bonney et al., 2016; Shirk et al., 2012), which not a few see as synonymous with community science. We disagree. CBC is *not* a community science project, yet not due to its co-created build up outside standard academia nor due to other criteria mentioned above.

It is no accident or contingency that the CBC does not typically address environmental, social or epistemic *injustice* even though it is an integral part of a highly activist environmental NGO (Shavit & Griesemer, 2018). Indeed, citizen science is not necessarily committed to seek environmental injustice, while community science is so committed. According to our account it cannot be labeled 'community science' not because of this fact, but because of its underlying working model framework. The CBC seeks volunteers qua *individuals*, i.e., as human beings in general, who primarily aim to solve a *universal* scientific question, such as mapping species presence (Silvertown, 2009).[7] CBC is obviously a valuable and important science-society research interaction but it is better characterized as 'citizen science'. It should be so labeled because it does *not necessarily* address the volunteer's local community (nor notice its particular heritage and current eco-social condition), *nor* her specific locality (the geographical scale size does not matter but its lack of particularity does). Following the same nonparticular approach, *it is* necessary for the CBC to aim for its data and metadata to be stored in a *universal* database (for example e-bird), ideally completely open yet in practice used mainly by affluent individuals and institutions.

Though it is typically "citizens" within a community who are involved in doing community science, our point is that their role is not based on their universal equal

[7] It is a yearly, 1-day bird count, done by tens of thousands of volunteers and used by professional scientists and by various conservationists. It began on Christmas 1900. Frank M. Chapman, an officer of the Audubon Society, called upon everyone to change their Christmas day ritual and count birds rather than hunt them. Twenty-seven bird watchers answered his call and many more are involved today. For details, see: https://www.audubon.org/conservation/science/christmas-bird-count/historychristmas-bird-count

rights as citizens, but rather on their attachment to a particular collective. Being a participant in the sense of being "part of" a collective requires them to explicitly notice—most likely also to compare—their inner and outer difference from other collectives, its structure and dynamics. Such attention to social heterogeneity, which is inherent to collectives (Shavit & Ellison, 2021), typically calls for an active response *action* if the differences revealed are unjustly hierarchical. In other words, the practical meaning of 'community science' is predisposed towards critical *joint action* regarding a particular collective via *collaboration* between *different* group members. According to Olson (1971), the goal of community science participants is to enhance their status as a whole and achieve a common group objective. Thus, community science is a social project, based on social learning, collective action and commitment to community goals (Loucks et al., 2017).

At its core, the split we suggest resides in the interaction-oriented framing of 'participation'. Participants of citizen science are invited to interact as "part of" the scientific endeavor without explicit reference to their particular collective backgrounds, difference from others, and without *attention to the interactions within and between them and others* regarding such differences. In contrast, community science intentionally and necessarily invokes one's agency and context of collective place, identity and belonging. This demarcation refers to a value-laden categorical difference alongside a continual difference in the degree of science democratization (Table 11.1). The categories in our model framework—collectivity, individuality, particularity and universality—are not only basic concepts per se but also represent epistemic values that guide actual research by foregrounding working practices. These practices shape and direct the science-society research interactions. Therefore, the aim of this table is not only to suggest semantic order to a fuzzy field, but also to notice which kind of values and practices are most relevant to which kind of science-society interaction.

One may think that a continuum lies between the opposites of 'collectivity' and 'individuality' or between 'particularity' and 'universality', yet this is not the case here. The opposite of a particular site, individual or group is a general, not a universal, one. Why? Because a general description abstracts away from some particular characteristics while importantly maintaining others, yet a universal description does something different. Universality is achievable via idealization rather than abstraction, i.e. by an attempt to *erase* all unique characteristics and *add* instead an idealized, in other words practically non-existent, traits that allow 'universality' to emerge. Similarly, 'collectivity' is not the opposite of 'individuality'. Why? Because a collective necessarily harbors kinds of interactions that a mere multiplicity of individuals need not (Griesemer & Shavit, 2023). In other words, at the opposite end of the continuum from 'individual' lies 'collection', not 'collective'. 'Individuality' and 'Collectivity' are different categories, leading to very different roles for 'interaction', as illustrated bellow.

The shift from one category to the next may carry profound implications in the context of critically addressing social, environmental and epistemic injustice. Community science, with its emphasis on the needs and aspirations of a particular community, emerges as a potent instrument for critically addressing, at times also actively reducing, injustice. By centering research efforts within a community, community science aligns its goals with that group's lived experiences and concerns. Such attention readily suggests the prioritization of localized questions and stronger intervention measures, which are typically avoided when a global perspective is employed, i.e., in citizen science research initiatives.

Overall, community science's commitment to address community-defined objectives is pre-disposed towards noticing and aiming to resist injustice. This happens because a collaborative ecosystem, where scientific interventions are tailored to a community's particular circumstances, is part of its practical meaning. Thus, community science may not only enhance the relevance and effectiveness of scientific research per se but also underscores its potential as a catalyst for critical change. In other words, 'science' becomes a justice-providing tool.

11.6 Community Science Substantially Involves Its Participants

We are not the only ones, nor the first, to argue for prioritizing community science. This holds not only in general but especially in peripheral localities, with and for under-represented groups, for the epistemic benefit of advancing science and science-based policy. Weisberg et al. (2023) have been conducting a long-term project of community science in the Galapagos in relation to climate change. They believe that projects using a community science model will likely have a greater impact on participants than projects with shallower engagement and can provide a good model for increasing environmental awareness and action.

In community science, the aim is to see the community not just as a participant but as a co-creator and co-owner of the research. This ownership fosters a sense of empowerment and agency among community members, leading them to more active and meaningful ways to affect the scientific process and its resulting policy recommendations, often to the benefit of innovative science as well as to more powerful resistance to injustice (Wylie, 2014). Given the above, community science projects tend to be more deeply embedded within the context of the community they serve. Researchers in community science projects prioritize issues that are relevant and significant to the community's wellbeing. This ensures that the research directly addresses the community's concerns, bringing more local figures to the fore, thus rendering participation more meaningful and impactful (Raviv & Silver, 2017).

Community science recognizes the expertise that community members bring to the table. It acknowledges that local knowledge and lived experiences are invaluable assets in the research process. As a result, community members will plausibly contribute their expertise, perspectives and insights more frequently. Conflicting and only partly overlapping ontologies and research methods are expected to emerge, actually enriching the study with a depth of understanding that external experts may lack (Ludwig, 2016). This collaborative approach enables research strategies that are not only evidence-based but also culturally sensitive and contextually appropriate (Ludwig, 2016).

Community science projects emphasize the establishment and strengthening of community ties. The collaborative research process can lead to increased social cohesion, trust and a sense of belonging within the community. This aspect of community building is often an essential part of community science initiatives. Community science projects, given their community scale of collaboration, typically require a sustained, long-term engagement with the complex social structure of a community as a whole. Researchers work closely with different community members over extended periods, building relationships of trust. This continuity allows for ongoing participation and the development of enduring partnerships. For enduring participation in a heterogenic system, it is especially relevant that dissent is *not* put aside for the sake of obtaining a quick and efficient consensus, but rather explicitly discussed (Longino, 2022; Shavit & Silver, under review). In the long run, such dialogic practices are expected to help produce more sustainable science-based policy.

In summary, community science is a deliberate and transformative approach that recognizes the community as an essential partner in the research process. This approach not only empowers the community in general but is especially relevant to peripheral communities, since it also enhances the relevance, impact and sustainability of the research outcomes.

11.7 Community Science and Resistance to Injustice

As community members engage in collaborative research efforts that directly address their surroundings, they tend to deepen their connection and respect for their own cultural heritage (Wylie, 2015). This connection is expected to foster a heightened sense of place (Haywood, 2014). This sense of belonging is not only rooted in the community's shared experiences but also in the active role that community members play in shaping their collective destiny. A positive sense of place plays a pivotal role in promoting social and environmental justice, particularly in peripheral areas where disparities and marginalization often prevail. It serves as the foundation upon which equitable development can be built. When communities have a strong and positive connection to their surroundings, it fosters a collective identity and pride that can serve as a powerful catalyst for social change (Chavis & Wandersman, 1990; Pretty, 2003). In peripheral regions, where social and economic capital and opportunities are typically limited, a robust sense of place can galvanize community members to advocate for their rights, demand fair allocation of resources and challenge systemic inequalities (Loucks et al., 2017). It encourages residents to take ownership of their environment and culture, ultimately leading to more inclusive decision-making processes and policies that prioritize the needs of the community.

In noticing local culture, it is easier to recognize that communities possess valuable local expertise. This knowledge is often overlooked in universal approaches of citizen science (Ottinger, 2013). Community science cannot but notice this knowledge, hence is more likely to leverage this expertise, enhancing the quality and appropriateness of research and interventions. Moreover, since community science acknowledges and respects the cultural nuances and values of the community, it is more sensitive to it. This sensitivity is critical for addressing social, environmental and epistemic injustices, as it increases the chance that interventions are culturally acceptable and do not inadvertently perpetuate harm.

In community science, there is a direct line of accountability between researchers and the community they are working with. This accountability can drive a stronger commitment to achieving outcomes that reduce or at least pro-actively resist injustice. Reducing injustice is also fostered by sustained, long-term engagement with the community because it enables strategic ongoing monitoring of progress/decline. Finally, working closely with a specific community over time helps sustain trustful relationships between researchers and community members. Universal approaches of science-society communication often struggle to establish such deep trust (Ankeny, under review; Lionelly, under review). Together with accountability, trust is essential for effective collaboration and for the implementation of changes aimed at reducing injustice.

As already mentioned, in community science, the community is an active participant and often a co-creator of solutions. This empowers community members to take ownership of the research and the actions that follow. When people have a say in the research process, they are more likely to be aligned with it, and more likely committed to implementing its recommendations (Pinto-Martin, 2021). Community science often involves community mobilization efforts. This can lead to increased social cohesion and collective action within the community, which can be a powerful force for reducing injustices. Research generated through community science is thus compelling to policymakers because it is grounded in the experiences and data of a specific community. This can lead more easily to policy changes that address injustices more effectively.

The universal appeal of science can also raise awareness of specific issues and mobilize resources more quickly and effectively, rendering it a *Modus operandi* for some goals (think about passing climate change resolutions in the UNIFFC). Community science, however, with its focus on a particular community, offers a more nuanced, context-specific, targeted and effective long-term approach to address the complex nature of injustice. The next section will empirically illustrate this point. We briefly present a 12-year project in a peripheral area ("Town Square Academia"), explicate its conceptual assumptions, track their different practical implementation as community science or citizen science activities, and unfold their different results.

11.8 An Illustrative Example: Town Square Academia

Town Square Academia (TSA) was founded in 2011, in the peripheral town of Kiryat Shmona, near the sources of the Jordan River, in the midst of a global social justice movement, exemplified by the "Arab Spring". The aim of TSA is to galvanize an interdisciplinary, activist and critical academia to use knowledge to connect and build a dialogue between heterogeneous groups within a peripheral region. In practice, each year over 1000 persons participate in 10 courses, 30 field-surveys and two live symposia, all free[8] and led by volunteering—academic and local—experts, who jointly attend a "Laboratory for Involved Academia".[9]

Participants come from different backgrounds (i.e., ethnicity, gender, age, religious sentiment, and socioeconomic level) and activities are conducted in various localities (riverbeds, hummus joints, schools, community centers, etc.). A TSA course, survey, monitoring program, or grant application must be built upon a *relevant* question to the actual life of a particular community—so that addressing that question may actively change something—yet the research process will not commence if that community is geographically, religiously or ethnically homogenous. TSA's goal is to conduct research with and for a particular, though heterogenous, community of communities.

Similarly, it is crucial that the research experts do not all deliver the same message, but rather foreground a critical dialogue between them and set aside ample time for public discussions. Therefore, TSA activities are co-led either by academics and local experts, academics educated in different disciplines, or by academics within the same discipline but holding different—often dissenting —views, all peer-reviewed. Practically, partnerships with new communities can be initiated in several ways—either by reaching out to community representatives or by responding to their inquiries. Our process follows a fairly consistent structure. It begins with a meeting, where we engage in open dialogue to explore the relevance and potential of joint activities. Together, we co-design the partnership's framework, whether it takes the form of a short course, an action group, or collaborative research.

During this first meeting, we identify relevant local and academic experts, as well as additional communities that should be involved in the process. In some cases, the meeting leads to an action plan that includes a coalition of organizations or communities that jointly shape and implement the initiative.

Occasionally, activities emerge following an approach from an academic expert. In such cases, we work together to determine the relevant local experts and additional academic collaborators, ensuring that diverse perspectives and knowledge

[8] Registration to a full TSA course (typically 5–6 2-hour meetings, though some are longer) costs 7$, and attending field-surveys and symposia are free of charge.

[9] The laboratory participants represent a highly diverse group with varied backgrounds. Religiously, they include Druze, Bedouins, and Jews with differing levels of observance. Professionally, they come from both academic and community sectors, bringing expertise from a range of disciplines. Additionally, the group includes representatives from national and local authorities, third-sector volunteers, and local residents, fostering a rich and multifaceted exchange of perspectives.

systems are incorporated. A guiding question throughout this process is: "*Who is not in the room?*" We actively seek to engage communities that reside within the region but have not yet participated in TSA's various initiatives. This is a grassroots work. In addition, participants (e.g., local activists, stakeholders or any community member) typically co-decide the subject-matter, goals, plans and methods with the scientists and/or TSA management team via roundtable discussions at the end of nearly every large or high-profile TSA activity.

This mode of science-society interaction reflects TSA's aim to resist epistemic injustice (Shavit et al., 2017), "a wrong done to someone specifically in their capacity as a knower" (Fricker, 2007, p. 1). Here it means not recognizing local expertise, memories and epistemic values of under-represented groups as relevant and potentially testable knowledge. Because of this motivation, and because of TSA's heterogeneous social composition and scale of operation, its knowledge-production processes can more easily increase the visibility of local knowers via face-to-face deliberation (Raviv & Silver, 2017). This emphasis on real-world relevance aligns with one of the project's central tenets: its pertinence to local environment. Throughout the years, various joint action frameworks have been implemented (e.g., courses, training trips, workshops, surveys, and symposia) with a primary focus on establishing long-term action groups.

One of these action groups, known as "זהב כחול" (Blue Gold), was established in 2014, on the banks of "עין זהב" (The Golden Spring) in Kiryat Shmona. The decision emerged at the end of a TSA course: "Some Things We Don't Know About Water". Course participants visited the spring to check water quality and wanted to continue and expand the work. Consequently, the group met, and decided to survey the stream beyond its aquatic dimension, connecting water monitoring to biodiversity conservation and the city's history, heritage, image and pride. The political situation in the region puts the city residences at a clear disadvantage compared to their neighboring Kibbutzim, who are situated near three other—much more famous—streams, benefit from a much stronger flow of tourism, a strong sense of place attachment (Lewicka, 2011), and their villages continuously grow while the city's population shrinks. During the TSA course, city residents talked about their local nature, local politics and the stream as part of their local identity. They thought that a stronger connection to local wildlife would be relevant in other dimensions as well.

Plant surveys began, with a local expert collecting information not only on species presence but also on the group's values and goals. During this process an endangered plant (*Verbascum galilaeum*)—which happens to resemble a *menorah* (candelabrum used in Jewish worship)—was marked by another local activist as a flagship species, rallying around it educational and community activities as well as academic seminars analyzing its distribution under different treatments of the city municipality. Surveys were pre-built so as to invite families with children, and all experts in leading positions—scientists and residences alike—explicitly choose to volunteer and use the available funds for other needs of the project. Additional surveys followed on other species, and gradually the group of pro-active community members grew from 5 to over 150.

These initiatives and their resulting meta-data were first stored in the TSA's course database by a local expert and participant. This later determined how our data and metadata would be collected, stored and used in our collaboration with Israel's Museum of Natural History and the Society for the Protection of Nature in Israel (SPNI). The museum and SPNI combined data-team developed the App "Bio-Collect", tailored to the particular interests of the peripheral "Blue Gold" group in a manner that will render its information interoperable in renowned international databases ("Living-Atlas"). Some of the meta-data fields tracked in this "Bio-Collect" version also fit any citizen science survey: time, location and the personal details of the specie's identifier. Yet some of the meta-data tracked by this application are categories that necessarily require interaction with a particular community, dynamically changing itself. For example, some of the data fields: traditional plant usage, special value (e.g., a flagship species chosen for its reflection of local tradition and as a catalyst for community agency), can only be found in a community science project. After the first course, a document describing the group's vision of the Golden Spring's future was co-written, in order to build long-term trustful collaboration, only to be dismissed soon after, as the local municipality's absence from the discussion was noticed and criticized by TSA's participants and leading experts. Since then, all activities at the stream aim to collaborate with the municipality and other governmental stakeholders, even when disagreements run deep.

Four years later, together with the city municipality and a governmental stakeholder (Lake Kinneret Drainage Authority), a grant for stream restoration was submitted, receiving 10 million NIS (about 3 million US$). Water and biodiversity surveys were re-opened and co-designed with academic experts from different disciplines, alongside local activists, local experts and municipality employees (e.g., the sewage system technician and the head of the business and entrepreneurship department). The sampling point locations, the parameters to be measured and how and when to measure them were determined through dialogue, which was often not easy for those experienced in conventional science. For example, a professor of water nanotechnology, who is extremely committed to science communication to the public, withdrew himself from the research due to disagreements with local experts. Changing one's sampling plan according to local considerations was too much for him, yet he continues to volunteer as an academic teaching expert in water quality courses, and continues to provide an outstanding example of academic volunteering. These repeated multidisciplinary and multiorganizational dialogues have ignited the emergence of a pressing new requirement: the establishment of a comprehensive research framework. Consequently, the "Involved Research Laboratory" was founded, serving as the nucleus for a collective research consortium comprised of dedicated researchers.[10] This collective body of researchers encompasses all stakeholders engaged in the research from its initiation, spanning from educators, third-sector and public-sector representatives, academic experts from the

[10] A 'researcher' is anyone conducting research in accord with acceptable academic norms, practices and testability barrier, and not only one holding a Ph.D. or professorship.

humanities, social and natural sciences, local experts, engaged residents, and other contributors. About 15 participants are bringing forth their unique expertise and knowledge, fostering a community-based collaborative research process.

11.9 Conclusion

We have shown that community science differs from citizen science in important ways. Not only semantics but also their resulting practices differentiate these two modes of science-society interaction; and not different degrees within the same category, but different categories explain this distinction. The particular versus the universal and the collective versus the individual are categories that readily notice different types of data and meta-data as casual difference makers, different sampling protocols as crucial for answering a question, different epistemic desiderata for the analytical model attempting an answer and, above all, a different interaction process for reaching a decision at each and every step of the way. In short, these semantic distinctions readily lead to substantially different research projects and policies. Given the ever-growing divergence and inequality between different communities in our era of seemingly flat, easy and global communication, and given the current importance of science and technology to determine the social and environmental fate of communities and societies, bringing scientists and community members together into a collective community of research seems like a worthy goal: epistemically, practically and ethically.

One seems justified in asking, is community science 'science' at all? Counting birds or galaxies in citizen science produces publishable data in scientific journals, but is community relevant data similarly publishable? Looking at the table shows it is located so far from basic or applicable science, one can reasonably ask: if the community holds very different research methods, and given that such methods are acknowledged in community science, are their resulting predictions testable at all? It should be noted that not every interaction between scientists and community members need be labeled 'community science'. For that, two requirements must be upheld. First, scientists need to join the community *qua* scientists, using their own norms of explanation and testability. Second, the community science project must be co-created and co-managed, so that all those involved have a say, including the scientists. In such cases, the testability bar must be passed and 'science'—as broadly construed in our opening section—must be the aim. Finally, if scientific publications attest to upholding standard science, then the 16 peer-reviewed papers and grants actually resulting from TSA seem to speak for themselves.

Acknowledgements We hold a deep sense of gratitude to the many people involved in this project and in Town Square Academia at large. Prof. Tali Tal is the thesis advisor of one of us, and without her valuable advice and support academic life would not have been possible. In addition, ever since the summer of 2011, numerous residences, local experts and academic experts have been volunteering their time and effort to "Town Square Academia". In particular, Sarit Kojikaro-Elimelech and Sivan Margalit from Kiryat Shmona alongside Prof. Giora Ritbow and Zohar Shief-Sinai from Tel Hai College, have shaped this project in their form. It is a true honor to work with such people. Finally, we deeply thank the support of the Ariane de Rothchild Women Doctoral Program and Yad Handiv Foundation (the "City of Streams" grant).

Appendix: Glossary

There are no universally accepted definitions for many of the concepts in our paper, therefore we have added this short and preliminary glossary:[11]

Citizen science—research activities involving academia-community collaboration, driven by individualistic and universal perspectives.

Community science—research activities conducted collaboratively between academia and the community, characterized by a dominant collectivist and particularistic perspective.

Democracy of knowledge is the degree of participation/involvement of citizens or communities in science (Dickinson & Bonney, 2017).

Epistemic Injustice is "a wrong done to someone specifically in their capacity as a knower" (Fricker, 2007).

Critical theory is a philosophical approach that critiques and seeks to change society by examining power structures, ideologies, and social inequalities. (Stanford Encyclopedia of Philosophy, 2023).

Science shop is a project in which nonscientists pose a problem in need of an answer (Urias et al., 2020).

Universal typically refers to something that is applicable to all cases or situations without exception, encompassing a widespread approach rather than being limited to specific instances or groups (Cambridge Dictionary, 2024).

Particular refers to something that is specific to a certain case, situation, or group. It emphasizes a focus on individual details, characteristics, or circumstances rather than a broad or general approach. In contrast to 'universal', which is all-encompassing, 'particular' is concerned with the unique aspects and specifics of a given instance or context (Cambridge Dictionary, 2024).

Individual: Pertaining to a single person [or other living entity], emphasizing personal goals, interests, and actions over those of a group (Oxford English Dictionary, 2023).

Collective: Pertaining to a group of interacting individuals holding joint goals, interests or actions in addition to their individual goals, interests and actions (Griesemer & Shavit, 2023; Shavit et al., 2017).

[11] We deeply thank an anonymous reviewer for this advise

References

Aharon-Gutman, M., Schaap, M., & Lederman, I. (2018). Social topography: Studying spatial inequality using a 3D regional model. *Journal of Rural Studies, 62*, 40–52.

Apple, M. W., & Weis, L. (1986). Seeing education relationally: The stratification of culture and people in the sociology of school knowledge. *Journal of Education, 168*(1), 7–34.

Arnstein, S. R. (1969). A ladder of citizen participation. *Journal of the American Institute of Planners, 35*(4), 216–224.

Berkes, F., & Folke, C. (Eds.). (1998). *Linking social and ecological systems: Management practices and social mechanisms for building resilience*. Cambridge University Press.

Berkes, F., & Ross, H. (2013). Community resilience: Toward an integrated approach. *Society & Natural Resources, 26*(1), 5–20.

Boersch-Supan, P. H., Trask, A. E., & Baillie, S. R. (2019). Robustness of simple avian population trend models for semi-structured citizen science data is species dependent. *Biological Conservation, 240*, 108286.

Bonney, R. (1996). Citizen science: A lab tradition. *Living Bird, 15*(4), 7–15.

Bonney, R., & Dickinson, J. L. (2012). Overview of citizen science. In J. L. Dickinson & R. Bonney (Eds.), *Citizen science: Public participation in environmental research* (pp. 19–26). Cornell University Press.

Bonney, R., Ballard, H., Jordan, R., McCallie, E., Phillips, T., Shirk, J. L., & Wilderman, C. C. (2009). *Public participation in scientific research: Defining the field and assessing its potential for informal science education* (A CAISE Inquiry Group Report (Issue July)). Center for Advancement of Informal Science Education (CAISE).

Bonney, R., Cooper, C., & Ballard, H. (2016). The theory and practice of citizen science: Launching a new journal. *Citizen Science: Theory and Practice, 1*(1), 1–1.

Butkevičienė, E., Skarlatidou, A., Balázs, B., Duží, B., Massetti, L., Tsampoulatidis, I., & Tauginienė, L. (2021). Citizen science case studies and their impacts on social innovation. In K. Vohland, A. M. Land-Zandstra, L. Ceccaroni, R. Lemmens, J. Perelló, M. Ponti, R. Samson, & K. Wagenknecht (Eds.), *The science of citizen science* (pp. 309–329). Springer.

Celikates, R., & Flynn, J. (2012). Critical theory. In E. N. Zalta & U. Nodelman (Eds.), *The Stanford encyclopedia of philosophy*. The Metaphysics Research Lab. Retrieved from https://plato.stanford.edu/entries/critical-theory/

Charles, A., Loucks, L., Berkes, F., & Armitage, D. (2020). Community science: A typology and its implications for governance of social-ecological systems. *Environmental Science & Policy, 106*, 77–86.

Chavis, D. M., & Wandersman, A. (1990). Sense of community in the urban environment: A catalyst for participation and community development. *American Journal of Community Psychology, 18*(1), 55–81.

Cooper, C. B., Lewenstein, B. V., Cavalier, D., & Kennedy, E. B. (2016). Two meanings of citizen science. *The Rightful Place of Science: Citizen Science, 2*, 51–62.

Dickel, S., & Franzen, M. (2016). The "problem of extension" revisited: New modes of digital participation in science. *Journal of Science Communication, 15*(1), 1–15.

Dickinson, J. L., & Bonney, R. (Eds.). (2017). *Citizen science: Public participation in environmental research*. Cornell University Press.

Dillon, J., Stevenson, R. B., & Wals, A. E. (2016). Introduction to the special section moving from citizen to civic science to address wicked conservation problems. Corrected by erratum 12844. *Conservation Biology, 30*(3), 450–455.

European Commission. (2014, January 21). *Green paper on Citizen Science for Europe: Towards a society of empowered citizens and enhanced research*. European Commission. Archived from the original on 17 October 2016. Retrieved 18 Aug 2016.

Fink, D., Johnston, A., Strimas-Mackey, M., Auer, T., Hochachka, W. M., Ligocki, S., ... & Rodewald, A. D. (2023). A double machine learning trend model for citizen science data. *Methods in Ecology and Evolution, 14*(9), 2435–2448.

Freire, P. (1970). *Cultural action for freedom*. Harvard Education Press.
Fricker, M. (2007). *Epistemic injustice: Power and the ethics of knowing*. Oxford University Press.
García, F. S., Pelacho, M., Woods, T., Fraisl, D., See, L., Haklay, M. M., & Arias, R. (2021). Finding what you need: A guide to citizen science guidelines. In *The science of citizen science* (p. 419). Springer International Publishing.
Giroux, H. (1983). Theories of reproduction and resistance in the new sociology of education: A critical analysis. *Harvard Educational Review, 53*(3), 257–293.
Griesemer, J., & Shavit, A. (2023). Scaffolding individuality: Coordination, cooperation, collaboration and community. *Philosophical Transactions of the Royal Society B, 378*(1872), 20210398.
Haklay, M. (2012). Citizen science and volunteered geographic information: Overview and typology of participation. *Crowdsourcing geographic knowledge: Volunteered geographic information (VGI) in theory and practice*, 105–122.
Haklay, M. E. (2015). *Citizen science and policy: A European perspective*. Woodrow Wilson International Center for Scholars.
Haklay, M. M., Dörler, D., Heigl, F., Manzoni, M., Hecker, S., & Vohland, K. (2021). What is citizen science? The challenges of definition. In *The science of citizen science* (p. 13). Springer.
Haywood, B. K. (2014). A "sense of place" in public participation in scientific research. *Science Education, 98*(1), 64–83.
Individual. (2023). *In Oxford English dictionary* (3rd ed.). Oxford University Press. https://www.oed.com/search/dictionary/?scope=Entries&q=individual
Irwin, A. (1995). *Citizen science: A study of people, expertise, and sustainable development*. Routledge.
Jordan, R., Gray, S., Sorensen, A., Newman, G., Mellor, D., Newman, G., Hmelo-Silver, C., LeDeau, S., Biehler, D., & Crall, A. (2016). Studying citizen science through adaptive management and learning feedbacks as mechanisms for improving conservation. *Conservation Biology, 30*(3), 487–495.
Jordan, R. C., Sorensen, A. E., & Gray, S. A. (2022). Participatory modeling in support of citizen science research. *Forests, 13*(4), 567.
Kelling, S., Fink, D., La Sorte, F. A., Johnston, A., Bruns, N. E., & Hochachka, W. M. (2015). Taking a 'Big Data' approach to data quality in a citizen science project. *Ambio, 44*, 601–611.
Kitcher, P. (2001). *Science, truth, and democracy*. Oxford University Press.
Krings, A., Kornberg, D., & Lane, E. (2019). Organizing under austerity: How residents' concerns became the flint water crisis. *Critical Sociology, 45*(4–5), 583–597.
Kuhn, T. (1962). *The structure of scientific revolutions*. Chicago University Press.
Lewicka, M. (2011). Place attachment: How far have we come in the last 40 years? *Journal of Environmental Psychology, 31*(3), 207–230.
Longino, H. E. (1990). *Science as social knowledge: Values and objectivity in scientific inquiry*. Princeton University Press.
Longino, H. E. (2022). What's social about social epistemology? *The Journal of Philosophy, 119*(4), 169–195.
Loucks, L., Berkes, F., Armitage, D., & Charles, A. (2017). Emergence of community science as a transformative process in Port Mouton Bay, Canada. In *Governing the coastal commons* (pp. 43–59). Routledge.
Ludwig, D. (2016). Overlapping ontologies and indigenous knowledge. From integration to ontological self-determination. *Studies in History and Philosophy of Science, 59*, 36–45.
Mgbeoji, I. (2007). *Global biopiracy: Patents, plants, and indigenous knowledge*. UBC Press.
National Academies of Sciences Engineering and Medicine. (2018). *Learning through citizen science: Enhancing opportunities by design*. National Academies Press.
Olson, M., Jr. (1971). *The logic of collective action: Public goods and the theory of groups, with a new preface and appendix* (Vol. 124). Harvard University Press.
Ottinger, G. (2013). Changing knowledge, local knowledge, and knowledge gaps: STS insights into procedural justice. *Science, Technology, & Human Values, 38*(2), 250270.

Particular. (2024). *Cambridge dictionary*. Cambridge University Press. Retrieved from https://dictionary.cambridge.org/

Peters, M. A., & Besley, T. (2019). Citizen science and post-normal science in a posttruth era: Democratising knowledge; socialising responsibility. *Educational Philosophy and Theory, 51*(13), 1293–1303.

Pinto-Martin, J. (2021). Community based participatory research in public health. In *Building a global network for community science: Workshop report*, May 24–25. Accessed 24 Sept 2023.

Pretty, J. (2003). Social capital and the collective management of resources. *Science, 302*(5652), 1912–1914.

Raviv, A., & Silver, Y. (2017). Unique community-learning model: "Town Square Academia". *International Journal of Current Research, 9*(8), 56106–56116.

Rosenthal, I. S., Byrnes, J. E., Cavanaugh, K. C., Bell, T. W., Harder, B., Haupt, A. J., ... & Trouille, L. (2018). Floating forests: Quantitative validation of citizen science data generated from consensus classifications. *arXiv preprint arXiv:1801.08522.*

Roy, H. E., Pocock, M. J. O., Preston, C. D., et al. (2012). *Understanding citizen science and environmental monitoring* (Final Report on behalf of UK-EOF). NERC Centre for Ecology & Hydrology and Natural History Museum.

Schrögel, P., & Kolleck, A. (2019). The many faces of participation in science. *Science & Technology Studies, 32*, 77–99.

Shavit, A., & Ellison, A. (2021). Diverse populations are conflated with heterogeneous collectives. *Philosophy of Science, 118*(10), 525–548.

Shavit, A., & Griesemer, J. (2009). There and back again, or the problem of locality in biodiversity surveys. *Philosophy of Science, 76*(3), 273–294.

Shavit, A., & Griesemer, J. (2010). Transforming objects into data: How minute technicalities of recording "species location" entrench a basic challenge for biodiversity. In *Science in the context of application* (pp. 169–193). Springer.

Shavit, A., & Griesemer, J. R. (2018). Science and sentiment: Grinnell's fact-based philosophy of biodiversity conservation. *Journal of the History of Biology, 51*(2), 283318.

Shavit, A., Kolumbus, A., & Silver, Y. (2017). Epistemic collectives, heterogeneity, and injustice: The case for Town Square Academia. *Landscapes of Collectivity in the Life Sciences, 20*, 199.

Shen, B. S. (1975). Science literacy and the public understanding of science. In: Day SB. *Communication of scientific information*. New York (NY): Karger Publishers.

Shirk, J. L., Ballard, H. L., Wilderman, C. C., Phillips, T., Wiggins, A., Jordan, R., McCallie, E., Minarchek, M., Lewenstein, B. V., Krasny, M. E., & Bonney, R. (2012). Public participation in scientific research: A framework for deliberate design. *Ecology and Society, 17*(2), 29.

Silvertown, J. (2009). A new dawn for citizen science. *Trends in Ecology & Evolution, 24*(9), 467–471.

Smallman, M. (2018). Science to the rescue or contingent progress? Comparing 10 years of public, expert and policy discourses on new and emerging science and technology in the United Kingdom. *Public Understanding of Science, 27(6)*, 655–673.

Stilgoe, J., Lock, S. J., & Wilsdon, J. (2014). Why should we promote public engagement with science?. *Public understanding of science, 23(1)*, 4–15.

Strasser, B. J., Baudry, J., Mahr, D., Sanchez, G., & Tancoigne, E. (2019). "Citizen science"? Rethinking science and public participation. *Science & Technology Studies, 32*(2), 52–76.

Urias, E., Vogels, F., Yalcin, S., Malagrida, R., Steinhaus, N., & Zweekhorst, M. (2020). A framework for science shop processes: Results of a modified Delphi study. *Futures, 123*, 102613.

Universal. (2024). *Cambridge dictionary*. Cambridge University Press. Retrieved from https://dictionary.cambridge.org/

Wandersman, A. (2003). Community science: Bridging the gap between science and practice with community-centered models. *American Journal of Community Psychology, 31*(3–4), 227–242.

Weisberg, M. (2021). Introduction. In *Building a global network for community science: Workshop report*, May 24–25. Accessed 24 Sept 2023.

Weisberg, D. S., Kovaka, K., Vaca, E., & Weisberg, M. (2023). LAVA-lobos: Raising environmental awareness through community science in the Galápagos Islands. *Citizen Science: Theory and Practice, 8*(1), 1.
Wilderman, C., Barron, A., & Imgrund, L. (2003). The ALLARM program: Growth, change and lessons learned. *The Volunteer Monitor, 15*(1), 1–4.
Wylie, A. (2012). Feminist philosophy of science: Standpoint matters. In *Proceedings and addresses of the American philosophical association* (Vol. 86, No. 2, pp. 47–76). American Philosophical Association.
Wylie, A. (2014). Community-based collaborative archaeology. In *Philosophy of social science: A new introduction* (pp. 68–82). Oxford University Press.
Wylie, A. (2015). A plurality of pluralisms: Collaborative practice in archaeology. In *Objectivity in science: New perspectives from science and technology studies* (pp. 189–210). Springer International Publishing.

Chapter 12
Epilogue

Dina Tsybulsky and Tali Tal

12.1 Reflecting on the Importance of Citizen Science for Biology and Environmental Education

This book explores citizen science from the perspective of its pedagogical potential. It demonstrates its pivotal role in enhancing educational outcomes and advancing scientific inquiry, specifically in the realms of biology and environmental education. Overall, the chapters confirm that citizen science initiatives can effectively engage learners, facilitate different types of reasoning, foster a deeper understanding of the nature of science and scientific concepts, and promote environmental citizenship. The empirical evidence underscores the transformative potential of involving learners in authentic scientific endeavors.

Citizen science holds significant potential for revolutionizing biology and environmental education by providing authentic, real-world research experiences that empower students to become active contributors to scientific knowledge and environmental agency. This potential can be observed in several key areas:

- **Fostering an Understanding of the Nature of Science**: Citizen science initiatives have the potential to foster a deeper understanding of the nature of science, a crucial aspect of scientific literacy (see, for example, **Ben Simon and Tsybulsky**, Chap. 7). By engaging in authentic scientific practices, students

D. Tsybulsky (✉)
Faculty of Education in Science and Technology, Technion Israel Institute of Technology, Haifa, Israel
e-mail: dinatsy@technion.ac.il

T. Tal
Faculty of Education in Science and Technology, Technion Israel Institute of Technology, Samuel Neaman Institute, Haifa, Israel
e-mail: rtal@ed.technion.ac.il

D. Tsybulsky, T. Tal (eds.), *Citizen Science Contributions to Biology and Environmental Education*, Contributions from Biology Education Research,
https://doi.org/10.1007/978-3-031-99298-8_12

learn that science is a dynamic and evolving process involving collaboration, uncertainty, and the constant refinement of knowledge based on evidence. This realization is essential for critically evaluating scientific claims, making informed decisions about socio-scientific issues, and participating in meaningful dialogue about the role of science in society.

- **Enhancing Understanding of Scientific Concepts**: Citizen science projects provide opportunities for authentic, real-world research experiences that enhance students' understanding of scientific concepts and processes. By participating in real scientific investigations, students gain firsthand experience with data collection, analysis, and interpretation that fosters a deeper understanding of scientific practices and their application to real-world issues. This practical, hands-on approach contrasts with traditional classroom learning, which often relies on textbook knowledge and hypothetical scenarios. This more engaging and meaningful learning experience further propels inquiry-based learning by shifting from standardized inquiry examples to the frequent reality of authentic research endeavors. These engage the students in actual (more uncertain) decision-making about tools, methods, and the type of data to collect and analyze. For example, **Lavie Alon et al.**, in Chap. 6, found that in a citizen science project that turned into a nature conservation effort, the students learned much more about rare species conservation, that boosted their interest and drive to protect heritage agriculture.
- **Fostering Environmental Citizenship**: Citizen science projects often focus on local environmental issues. These foster a sense of place and connection to nature, which are crucial for promoting environmental citizenship. By engaging in research that directly impacts their communities and surroundings, students develop a deeper appreciation for the importance of environmental agency and are more likely to take steps later on to address environmental challenges. This connection to the local environment increases the relevance and meaningfulness of scientific learning, and motivates students to apply their knowledge to real-world situations and become more active players in environmental conservation efforts (see **Part III** of this book, Chaps. 9, 10, and 11).
- **Promoting Collaboration and Mentorship**: Citizen science encourages collaborations between students, teachers, and scientists by breaking down the traditional barriers between scientific research and education. This partnership scaffolds a more inclusive and participatory approach to science, and can empower students to contribute to scientific knowledge, which gives them a powerful sense of ownership over their learning. Working alongside scientists provides students with valuable mentorship opportunities that may inspire them to pursue careers in STEM fields and become future leaders in scientific research and environmental conservation.
- **Addressing Scientific and Educational Challenges**: Citizen science can also help address challenges faced by scientists, such as limited resources and the need for large-scale data collection, while simultaneously enhancing the educational experience for students. By engaging the public in scientific research, scientists can access a larger pool of volunteers for data collection and analysis, which enables them to expand the scope and impact of their work. This mutually

beneficial partnership allows for the collection of more robust and geographically diverse datasets, leading to a more comprehensive understanding of complex environmental issues.

Overall, citizen science emerges as a powerful educational tool that not only enriches the learning experience but also contributes to the advancement of scientific knowledge and environmental agency. As educators and researchers, embracing the potential of citizen science can inspire the next generation to engage with science in meaningful ways, thus ultimately fostering a more scientifically literate and environmentally conscious society.

12.2 Key Factors for Successful Citizen Science School Projects in Biological and Environmental Education

The success of **citizen science** school projects depends on several key factors that can lead to more effective implementation and meaningful engagement. As shown in this volume, these critical elements and strategies include:

- **Educational leadership**: Teachers who show enthusiasm and take initiative in designing and implementing these projects are more likely to elicit support from the school administration, fellow teachers, and parents. This leadership is crucial given the collaborative nature of **citizen science**, which often involves coordination with scientists, community members, and school administration.
- **Selecting a locally relevant topic**: Choosing a research question or a problem pertinent to the school's local environment can generate student interest and a sense of purpose. Teachers should collaborate with local experts or community representatives to identify topics in citizen science that are both scientifically valuable and relevant to student inquiry.
- **Well-defined curricula**: Successful **citizen science** school projects require the design of curricula that integrate **citizen science** principles and scientific practices. Teachers need a solid background in the topic to guide curriculum development, and select and implement appropriate data collection and analysis methods. Creating engaging learning sequences, using visual aids, and encouraging reflection are essential aspects of an effective citizen science curriculum.
- **Assessing impact**: Evaluating the impact of citizen science projects on student knowledge, attitudes, dispositions, and motivation, as well as their contribution to scientific knowledge, is essential. This comprehensive assessment can help educators identify areas for improvement and judge the value of student participation in authentic research. These methods should include pre-and post-project questionnaires and interviews, student reflections, and the analysis of student work.
- **Flexibility and adaptability**: Flexibility and adaptability are crucial for managing the dynamic nature of **citizen science** projects. Teachers need to be prepared to handle unforeseen challenges, changes in the field, or the need to modify research questions. Adapting lesson plans and embracing unexpected developments in the research process are key to maintaining student engagement.

- **Co-creation and collaboration**: Effective **citizen science** projects are often the result of close collaboration between scientists and teachers. This partnership ensures that projects meet both educational and scientific goals. Educators bring their pedagogical expertise which makes it possible to align projects with curriculum standards as well as student needs and expectations. Scientists provide scientific guidance to ensure the quality and robustness of the research protocols.
- **Facilitating organizations**: Institutions such as universities or research centers, such as the Taking Citizen Science to School (TCSS)[1] center described in several chapters can play an important role in supporting **citizen science** projects in schools. These organizations can provide in-service professional development programs, contribute to project design, allocate resources, and connect schools with scientists. They can also help bridge the gap between scientific and educational communities by streamlining communication and coordination among the stakeholders.

By incorporating these factors, educators will be in a better position to design and implement **citizen science** projects that foster scientific literacy, environmental stewardship, and a sense of agency among students.

12.3 Hurdles to Implementing Citizen Science in Biological and Environmental Education and Optimal Solution Strategies

Although citizen science initiatives constitute a promising avenue for promoting environmental citizenship and engaging students in authentic scientific research, their successful implementation faces a number of hurdles. This book highlights a number of obstacles and suggests strategies for coping with them:

- **The complexity of citizen science initiatives: The multifaceted nature of citizen science** initiatives that encompass scientific research, education, and community engagement means it is often difficult to accommodate the often divergent goals and perspectives of each stakeholder. A clear strategy from the outset can serve this purpose if it outlines respective roles, responsibilities and communication channels. Designating a core team with joint expertise in science, education, and community engagement can help find a common language and coordinate efforts.
- **Teacher constraints**: Teachers who need to cover a comprehensive curriculum and prepare students for standardized tests may find it hard to develop or engage in citizen science projects. A lack of resources and few opportunities for specific professional development opportunities in dedicated groups or through institutional channels, all conspire to impede integrating citizen science projects into

[1] https://www.tcss.center/home-en

existing curricula. These well-known issues can be overcome to some extent by enabling teachers to have access to curriculum-aligned lesson plans, data collection protocols, and assessment tools. These considerably streamline the implementation process, reduce teacher workload and make citizen science more accessible and feasible within the constraints of the school day. By aligning citizen science projects with curriculum standards and providing clear connections to mandatory learning objectives, educators can ensure that these projects complement rather than compete with other educational priorities.

Teachers may feel overwhelmed by the prospect of incorporating a new pedagogical approach into their already demanding schedules, especially if they lack confidence in their own scientific backgrounds or experience with outdoor learning. To address these challenges, several strategies can be adopted (see **Golumbic**, Chap. 2; **de Groot and Boeve-de Pauw**, Chap. 8):

- **Providing teachers with professional development opportunities**: Training sessions and workshops can give these teachers the background and skills to implement citizen science projects in their classrooms confidently.
- **Developing curriculum-aligned resources**: Citizen science initiatives that align with existing educational standards and provide teachers with ready-to-use materials can significantly reduce the burden of curriculum adaptation.
- **Fostering collaborations between scientists and educators**: Partnerships between scientists and teachers can bridge the gap between research and education. This will help ensure that citizen science projects are both scientifically rigorous and pedagogically sound.
- **Scientific rigor vs. pedagogical goals**: Scientists leading citizen science projects often find it hard to adapt their research to educational settings and the level of scientific rigor to pedagogical goals. Establishing collaborations between scientists and teachers to co-create citizen science projects can help overcome this very real issue. This partnership can allow for the development of projects that are both scientifically sound and tailored to the specific learning objectives of the curriculum and age group of the students. It also ensures that the data collected by students are useful to the scientific community, which in turn can enhance the authenticity of the learning experience for the students.
- **Sustained engagement**: Effective citizen science initiatives require sustained engagement and communication among stakeholders, which can be hard to maintain over time. To foster more lasting partnerships, clear communication channels and opportunities for regular interaction between scientists, teachers, and students should be planned for and implemented. Joint workshops, field trips, and data analysis sessions can promote collaboration and strengthen relationships. In addition, online platforms can simplify and harmonize communication, data sharing, and ongoing project updates.
- **Recognition and support**: The lack of recognition and support for citizen science in educational system as a whole can hinder its widespread adoption. Efforts to raise awareness and advocacy are needed to overcome this systemic challenge. Highlighting the educational benefits of citizen science, such as greater scientific

literacy, critical thinking, and problem-solving skills, can garner support from policymakers and school administrators. Sharing successful case studies and demonstrating the positive impact of citizen science on student learning and the resolution of community issues can further encourage its integration into the education system.

By proactively addressing these challenges and implementing these solutions, citizen science initiatives can better bridge the gap between science and society, and empower students to become informed and engaged citizens.

12.4 Empowering Citizens to Become 'Agents of Change' through Citizen Science

Citizen science can empower citizens to become 'agents of change,' particularly in the context of environmental issues. This empowerment can be more successfully nurtured by a citizen-centric approach than a purely solely science-centric one and is crucial to achieving this goal. Several key strategies are outlined in this volume:

- **Co-designing Citizen Science Initiatives**: Involving citizens, including students, in co-designing environmental citizen science initiatives can be transformative (see, for example, **Lavie Alon et al.**, Chap. 6). This participatory approach empowers them to shape research questions, methodologies, and data analysis, which means that the initiative is more likely to align with their concerns and priorities.
- **Moving Beyond Data Collection**: For citizens to become agents of change, their participation needs to move beyond data collection (see, for example, **Yael Bar Nes et al.**, Chap. 3). It is important to engage citizens in the entire data life cycle—from co-designing the project to analyzing data, interpreting the findings, and at times taking action or taking part in policy and decision-making based on the insights gained.
- **Fostering Environmental Citizenship Competences**: Effective citizen science initiatives should aim to foster environmental citizenship competences, including critical thinking, problem-solving, collaboration, and communication skills. Through participation, citizens can develop a deeper understanding of environmental issues, learn how to use scientific evidence to advocate for change, and gain the confidence to take action.
- **Transformative Learning and Action Taking**: Citizen science can encourage transformative learning, a key element in empowering citizens as agents of change. This type of learning goes beyond acquiring knowledge because it can lead to shifts in perspectives, values, and behaviors. Through citizen science, individuals can develop a stronger sense of environmental responsibility, which may prompt them to engage in individual and collective actions in their personal and public lives.

- **CS4EC Framework**: The Citizen Science for Environmental Citizenship (CS4EC) framework is a valuable guide for designing citizen science projects that empower citizens (see **Andreas Ch. Hadjichambis et al.,** Chap. 4). This framework takes a three-phase approach:
  - **Citizens (including students) as co-designers**: Involving citizens in the initial design phase to ensure their concerns and priorities are addressed.
  - **Citizens as scientists**: Engaging citizens in scientific inquiry, data collection, and analysis.
  - **Citizens as agents of change**: Empowering citizens to take informed action and advocate for change based on their scientific findings.
  - When designed with a citizen-centric focus, citizen science can empower individuals to become active participants in addressing environmental challenges and driving positive change within their communities and beyond.

12.5 Future Research Directions

Looking ahead, there are vast opportunities for further research and development within the realm of citizen science education. Future investigations should explore the long-term impact on participants, innovative technological solutions for data collection and analysis, and the expansion of citizen science initiatives to underserved populations. The continuing evolution of this field should result in exciting advances and a fuller integration of citizen science into both educational and research contexts. Below are several potential future research directions related to citizen science and its role in biology and environmental education:

- Explore the potential of citizen science in addressing social and environmental injustices, particularly in marginalized communities. **Silver and Shavit**, in Chap. 11, suggested that community science, because of its focus on local needs and participatory nature, could be instrumental in addressing injustices. This points to a promising research direction that would investigate how citizen science projects can be designed and implemented to empower marginalized communities, amplify their voices, and contribute to more equitable and just outcomes. This type of study could examine features such as community engagement strategies, power dynamics within projects, and the long-term impact on community well-being and social change.
- Examine the effectiveness of different educational designs and interventions aimed at maximizing the pedagogical benefits of citizen science data in biology and environmental education. **Dvir and Tsybulsky,** in Chap. 5, emphasized the importance of well-structured educational interventions in maximizing learning outcomes. Future research could focus on developing, implementing, and evaluating innovative pedagogical approaches that leverage the richness of citizen science data. This could involve exploring different learning activities, scaf-

folding strategies, and assessment methods tailored to the specific challenges and opportunities presented by citizen science data.

- Investigate the interplay between different types of reasoning fostered by citizen science projects. While **Ben Simon and Tsybulsky,** in Chap. 7, highlight research indicating that citizen science provides a suitable pedagogical framework for developing scientific reasoning, statistical reasoning, and understanding the nature of science, they also suggest the need for further exploration of the potential relationships between these types of reasoning. For instance, does the development of one type of reasoning encourage or facilitate the development of others? Investigating these interrelationships could lead to more holistic and integrated approaches to citizen science education.
- Investigate the long-term impact of citizen science activities on students' scientific literacy, environmental agency, and self-efficacy. **Golumbic**, in Chap. 2, noted that participation in citizen science projects can lead to learning outcomes extending beyond content knowledge, such as stewardship and self-efficacy. Longitudinal studies could provide valuable insights into the lasting effects of these experiences on students' attitudes, behaviors, and career trajectories, which could ultimately inform the design of more impactful citizen science programs.
- Examine the factors contributing to the sustainability and longevity of citizen science projects, especially in the context of limited funding and volunteer engagement. **Gálvez et al.**, in Chap. 10, highlight the problems involved in sustaining citizen science projects, particularly as regards securing long-term funding and maintaining participant motivation. Research exploring innovative funding models, volunteer management strategies, and approaches for fostering long-term community engagement could be beneficial for ensuring the continued success and impact of citizen science initiatives. These research directions constitute a starting point for further exploration of citizen science and its multifaceted implications for scientific research, education, and society.

12.6 Final Thoughts

While compiling this book, we were struck by the key role that citizen science can play in both educational and scientific contexts, particularly in the fields of biology and environmental education. The research presented in this volume demonstrates the transformative potential of engaging students in authentic scientific investigations. By bringing together research from diverse contexts, this volume offers a unique synthesis of perspectives, highlighting both the challenges and opportunities of citizen science in biology and environmental education. Citizen science initiatives not only enhance students' understanding of biological and environmental principles but also foster critical thinking, environmental agency, and interdisciplinary collaboration.

Our journey has reinforced our belief in the value of integrating citizen science into educational frameworks. By involving students in real-world research, educators and scientists can bridge the gap between theoretical knowledge and practical application, thus making the study of biology and the environment more tangible and relevant. This approach not only enriches the educational experience but also cultivates the growth of a new generation of informed and active contributors to scientific and environmental communities.

As we look toward the future, we are optimistic about the continued expansion and deeper integration of citizen science into biology and environmental education, emphasizing its role in fostering scientific literacy and societal engagement. The evidence provided in this book highlights the substantial benefits and far-reaching implications of these initiatives. We are confident that the continued exploration and implementation of citizen science projects will inspire educators, researchers, and students alike, and lead to a more scientifically literate and environmentally conscious society. It is our hope that this book will serve as a foundation for future research and practice, by encouraging further engagement with citizen science and its profound educational potential in the realms of biology and environmental science.

Acknowledgments We would like to express our deepest gratitude to all those who contributed to this book. First and foremost, we thank the researchers, educators, and students who participated in the citizen science projects and shared their experiences and findings. Your contributions were integral to the success of this volume.

Special thanks go to our academic partners, who provided support, insight, and guidance throughout the research and writing process. We are grateful to the supporting organizations and funding bodies that financed some of the projects and studies included in this book.

We also extend our heartfelt thanks to the ERIDOB [European Researchers in Didactic Biology] community. Your continuous support, collaboration, and shared passion for biology education were invaluable in shaping the perspectives presented in this book.

Finally, we wish to express our sincere appreciation to our colleagues and families for their support and understanding during the writing and editing phases. Without your patience and encouragement, this work would never have been brought to fruition.

MIX
Papier aus verantwortungsvollen Quellen
Paper from responsible sources
FSC® C105338

If you have any concerns about our products,
you can contact us on
ProductSafety@springernature.com

In case Publisher is established outside the EU,
the EU authorized representative is:
Springer Nature Customer Service Center GmbH
Europaplatz 3, 69115 Heidelberg, Germany

Printed by Libri Plureos GmbH
in Hamburg, Germany